For more years than I want to record, I've been thinking about writing a book of some of my memories but as I look back today I wonder if these things are as important as I thought they were some years past when we would be telling friends some of the experiences and things that I have done, and later my wife and I would often her the comment "you should write a book." Well maybe, but I told some of these happenings so often, they seem to be old hat; however with pressure from my wife and family, I will endeavor to record some of my life and my families experiences through the years.

I was born early in life at about 8 ½ pounds in early spring and wanted to get a good start in life, however I was born on a Saturday rather late in the week, from then on as soon as I was able to understand, I was told to work hard for a living.

My mother and dad were married in Dorion, a small community in northwestern Ontario. My mother had come from Delhi Ontario and my dad from Wolverhampton England. When I was born, my mother was close to 25 and my dad was 31 years. My mother's ancestors came to North America in the early 1600's from Holland. Her maiden name was Vanderburgh, and some of her family came to Canada in the 1800's. My dad came to Canada with his two brothers in 1903 and they stayed or worked around eastern Ontario then decided to come further west where they took up three homesteads in Dorion 45 miles east of Port Arthur Ont., and after they were settled, they brought their sister to Canada. Their grandfather William came to Canada's and the U.S.'s west coast in 1793. He was the captain of the ship Chatham along with Captain Vancouver on the ship Discovery. Captain Broughton with his ship and crew spent the latter part of the winter in the mouth of the

Columbia River. While there, named a number of places. The spring and summer of 1793, Captains Vancouver and Broughton mapped and named most of the west coast of Canada.

Our home was on a farm that my dad had homesteaded when he came to Canada with two brothers in 1903. It was commonly known as the valley farm as it was in the Cold Water Valley. The river ran right below our house which was up on a hill. It was a nice frame home 1 ½ storeys with 2 bedrooms up and living room and kitchen on the main floor and also had a dugout root cellar under the living room. This cellar I remember at one Christmas when I was about two my parents bought me a toy dog that barked. I never did care much for it. One time when my mother was down in the cellar, the door was open, I threw the dog down and when my mother came back up she asked me why I had done that and I told her it was where it belonged as it made too much noise barking.

Another cellar experience I recall, we had been visiting neighbors, had supper there and didn't have any potatoes with the meal and on the way home I told mother and dad I was hungry as they didn't have any potatoes. When we arrived home, dad went down in the cellar and brought up a potato which mother peeled and I ate it raw and was contented as I had my potato.

While living on the farm, my dad caught a young crow and it became quite a pet for some time, then one fall it left with a flock of crows and I don't remember seeing it again.

When I was about 3 or 4 we got a scotch collie dog which was well trained bringing in the cattle. Dad would go out in the yard to call the cows and if they didn't come right away, he would send the dog to fetch them, so it didn't take long before the cattle knew they didn't want the dog after them and they would come home on their own. I remember being in the barn watching the dog putting the cows in their proper stalls, he was a real smart dog. We called him Colly. If there were friends who came to visit, he would watch my brother and me to make certain that no one touched us.

My brother Wray was born two years after me and he was the good looking one, had light curly hair and fine features. Mother wanted a girl and dad said as he was a boy he should look like one and was never too fussy about the light curly hair, and sometimes my mother would do his hair up in ringlets to tease my dad I thought. One time in the spring of the year we were out with dad while he was out working around the farm and I had my hands all sticky from black poplar brush, so I asked dad where I could wipe my hands, he said probably in my brothers hair, so that's what I did and his hair was a real stuck up mess and when he came into the house, mother wanted to know what had happened to his hair. While she tried to clean it, it wouldn't wash out and finally she used kerosene and part of his hair fell out and dad and I were in real trouble. Finally when his hair started to grow back, it was straight black, now dad said he looks like a real boy, but I'm not sure if mother ever forgave dad or me.

My mother was a real loving person, and I remember her, morning and evening sitting on my brother's or my bed telling us bible or other stories almost every day while living in that home.

There I have lots of memories of that home and farm, although I was only 6 years old when we moved to the center of the community so I would be closer to school.

I remember some of the things that happened before I was six. Dad would always go out and shoot a deer for meat supply, occasionally he would get a moose, however our dog Colly would like to go with dad hunting. Colly would go out and find a deer then would keep it from running away by nipping it on the hind legs, and the deer would turn to attack the dog then turn to leave and Colly would nip it again. While this was going on, there would be continuous barking so dad would know where he was. Sometimes he would pick a deer dad didn't want and dad would have to call him off much to Colly's disgust. Dad said that it took the pleasure out of hunting, and one time he locked the dog in the house and he jumped through a closed window and broke the glass as he knew dad had gone with the rifle.

Sometimes dad would be talking with neighbors or friends about some of the experiences his brothers and he had when they first moved on their homestead. They decided to build a temporary log shelter until they could build a house. It took quite a while to build a house with lumber as it had to be ripped with two men on what is called a head frame, which was a pit deep enough for a man to stand in, logs across the hole to hold the log that they were ripping. One man on the bottom would pull the rip saw down and one man on top would pull the saw up. The rip saw was about 6 feet long with a handle on each end. Getting back to the temporary shelter, when they started falling trees, they would fall a tree like a beaver and never knew where the tree was going to fall.

When they had some logs felled, they picked a place on a hill, then leveled a spot for the shelter and dug a trench on the upper side of the shelter so the water wouldn't run in, then they proceeded to build the shelter. When it came to the roof, dad's brother Percy, who had been in Australia saw some brush that he thought was water proof as he had used it there, so they put some poles on the roof to hold the brush up then covered the poles with brush. They got the shelter finished then piled their supplies inside. That evening it started to rain, dripping through the brush roof and while they were wondering what to do to keep dry, the trench at the rear of the shelter was filling with water, and all of a sudden the trench broke and the water came rushing through and around the shanty. They were wet and everything was wet. They did their best to keep warm for the night. The next day the sun came out and they had all their supplies and clothing hanging round on trees to dry out and of course they had to find something else to use on the roof. They had lost part of their food supply. They eventually built a nice log house with ripped lumber nailed to poles and cedar shakes on top of this for the roof. I remember being in the old house at night when I was a boy, looking out at the stars but it never leaked, I guess it must have been the way the shakes were installed. We used that old house for storage and a granary, as my dad and some neighbors built the house we lived in before mother and dad were married.

Dad always liked to pull a prank if he could, especially telling stories. One Christmas when I was about 4 years and my brother was 2, we had the Christmas tree all decorated and ready for Santa to arrive. Just before we were going to bed, my dad went outside and brought in a large wolf trap and said to us boys "now how would it be if we just set this trap here by the tree and when Santa comes, we will catch him and have his whole bag of toys for ourselves." well, I didn't think it was the right thing to do, and when no one was looking, I grabbed the trap and threw it out the back door. Mother and dad said later I must have really thrown it as they never did find the trap again.

As I got a little older, we would go down by the river and fish with dad and we would catch real nice trout. I fished on that river whenever I had any spare time right up to the time I left the area. It always was a nice quiet and peaceful area to spend an afternoon or evening.

One time when I was about three, I was kicked by one of the work horses on the left side of my chest and had the shoe print for quite a while. In the spring and fall when the horses were not working, dad would let them pasture the grass around the buildings, but would try to keep them away from the house, and occasionally would pick up a chip of wood and throw at the horses. Well I thought I could when no one was watching, and I didn't realize that the horse would kick, which he did and I was knocked out, almost frightened my mother to death, however, dad or mother revived me and covered my wounds and gave me a good talking to about never throwing anything at the horses.

We would go to the Baptist Church with a horse and buggy or a light sleigh in the winter. One time we were going out in the sleigh, dad tapped the horse on the rump with a line and the horse switched his tail and a hair hit me in the eye, broke a blood vessel in my eye and I have that scar today.

Going back, some of the stories dad would tell of experiences, I recall one of our neighbors dad hunted with was an excellent marksman, the same as my dad; I heard them talking about one experience they had when they were out hunting for moose. D.K. saw a moose and shot it, however the moose whose head was looking through between two large birch trees appeared in a standing position so he shot it again and the moose didn't move. When dad came to where D.K. was standing looking at his rifle and pointed at the moose still looking through the trees, he said to dad "you shoot the moose, my rifle or my eye sight has something wrong as I've never missed before." Well before dad took a shot, they decided to see how close they could walk up to it. As it turned out, they walked right up to it and it was dead. Only one problem. The moose had sat down with its head and horns caught between the trees on one side and the rest of the body on the other; they had to cut the head off before they could move the moose.

Before going further, I must tell you about dad's experience at baking bread. He was the youngest of three brothers, so the older two said dad had to be the cook. He had seen some cooking in England bur never had tried it. One day when they were getting groceries, he ordered yeast cakes, so decided to use them the same way as quick yeast that he had seen in England, so he started to make bread. After mixing, he put it out to rise, but it shrunk, so he thought maybe it had been too cool, so he put it in the oven and as they were cooking, the loaves got smaller.

Eventually they came out of the oven and they thought they would try some fresh bread; however they couldn't cut or break it. I think they did get some of the edges but decided dad had better find out more about baking bread. That same evening, one of the members of the community called in. The three brothers had eaten; however, they said they had this loaf of bread he was welcome to if he could eat it. As he was hungry he assured them he could as he had a good hunting knife. Well, he got the knife in one loaf and finally broke his knife trying to get it out and they got him something else to eat after they had their fun with the neighbour and the bread. The next day dad threw the bread over the bank down towards the river. About 15 years later dad was down along the river cutting some brush and hit his axe on something like a rock. When he dug in the mud, he found one of those loaves of bread that had petrified, so he took it up to the house and we used it to keep the door open in warm weather; if it wasn't a loaf of bread, it certainly looked like one. I should add that dad did become one of the top cooks in the area. He worked on the survey and laying of the steel on the Canadian National Railroad through northwestern Ontario and was top cook for the railroad by the time he was 19.

While we still live on the valley farm, I remember dad taking me hunting with him, of course that usually wasn't very far from home, and in this case it was to the top fields where the deer would feed in the evening. This time I guess I was probably 4 years old, maybe 5. Well any way, dad shot a deer and it was quite heavy, so he let me carry his rifle home after he took the ammunition out of it. At the time, I think he had a 32 Special which was a very small rifle, however after carrying the rifle a short distance it became too heavy for me so I hung onto the barrel and dragged it along of course behind my dad, as he wouldn't appreciate it being dragged.

Dad told about a fellow and his wife that moved close to them shortly after dad and his two brothers homesteaded. His name was Walford who had no experience at farming or living in the country. Dad said they had quite a few hilarious experiences with Walford. He had an ox but only had a yoke for two oxen, not knowing how to hitch the ox single. One day when he had some work to do with the ox, he got on the other side of the yoke with the ox, however the ox decided to have a little fun and Walford had to run too, and as it became more serious, he started to holler "help us help us we're running away." As this was close to where dad was working that day, he went over to see what he could do and stopped the runaway. Walford had become tangled in the harness and couldn't stop the ox, so dad showed him how to use his ox single; this was only one experience. When dad took Walford hunting, he would get lost as soon as he couldn't see his house. One time dad asked him if he ever used a compass, and he said he had, so he loaned Walford his compass and they went hunting, however it didn't have the house marked on the compass and that concerned him. Needless to say he didn't know anything about a compass, but he was a real jovial fellow so dad would take him along for laughs, and of course try to help him out.

In the spring when I was six years of age, my parents moved to the central part of the community so we would be closer to school. My uncle had a country store where almost everything was sold also a large farm which my dad took over, managed and farmed with one or two hired hands. We lived in a small two storey house on the farm. The farm had about 50 Holsteins cows, several horses and quite a few pigs. One pig I remember to this day, he almost caught me in the barn one day. Dad had told my brother and me to stay out of the pig barn; however I wanted to see this large black and white boar that was there. One day when dad was feeding, I came in behind and was looking at the size of the boar and also at the large tusks he had, when all of a sudden he jumped over his feed trough and ran straight at me. Well I really ran out in the barn yard but he was gaining on me and dad was running and yelling to run. I made it to the gate which I climbed over just in time. That boar missed my foot and leg by about two seconds. I never went back in that area again. After climbing over the gate, I went in the horse barn and up in the second storey overlooking the yard where the boar had chased me. Dad had quite a time getting him back in his pen, and really gave me a tongue lashing for being in there, especially after being told to stay out, however curiosity almost got me that time.

There were good large two storey horse barns and a cow barn and the farm work was all done with horses. There were large fields and there was a lot of milk shipped and had a large cooler room for milk storage; large hay lofts and hay barns, a great place to play summer and winter. I had other children and cousins to play with and explore with, and some were girls but that didn't matter as they were what was known as tom boys in those days.

As my 2nd brother got old enough to navigate, it took the whole family to watch him. He would climb over anything to get to where dad was, and of course he didn't help when you were trying to do farm work.

One year after Lorne was about two and he was trying to talk, while at times was rather crooked or backward to say the least; dad had a Dane who had come from Denmark right to our farm and couldn't talk any English, and when he was able to use some English, he would say Lorne was trying to talk his language and that was why we couldn't understand Lorne. August, the Dane was a big man and very strong, and he learned to talk the English language by using the English and Danish bibles. He was with us for about seven months and when he left he talked real good English. He went out to B.C. Okanogan and was going to fruit farm. We never did hear from him again.

During the summer of 1924 my first sister was born, she was a cute chubby little girl, and quite contented and happy. In September I was now seven and could start to school; the school was quite modern, a brick building, two classrooms, full basement and in the top of the building was a community hall. In later years I attended quite a few country dances there, of course there were Christmas concerts, fall fairs and other community affairs. Later years the hall was also used for a classroom. As the community grew, I took my 9th and 10th grades there, we had three teachers then, one acted as principal also. The school was noted for the tubular type of fire escape - it was a pipe you slid down. There was about 130 students and could empty the school in less than 2 ½ minutes. When I first started school, I had to walk about a mile, but along the road there were two farms which had geese and they would chase me almost every morning, eventually someone told me to carry a stick and hit them. The geese finally got the message and didn't bother me any more.

While we lived in this area, one of my aunts used to take me to the Anglican Church and Sunday school which I appreciated, later I returned to the Baptist Church.

We lost our dog Colly while living there. The country road ran right close to our house and he thought he should keep everything away from the house and of course he didn't know that cars couldn't stop for him, so one day was hit and killed on the road by a car.

One summer while we lived in that area, there was lots of excitement about a neighbors Holstein bull that was running loose and chasing the occasional party along the highway that ran through the community. Apparently the Ontario Hydro had put an electric power line through the neighbor's farm against his will and left some fences down, so the farmer wouldn't repair the fence. So needless to say his cattle bothered everyone but worst of all was the bull weighed 2000 pounds. One evening he arrived at our farm and was trying to break into our bull pen to fight with our bull which was about the same size. August, the Dane took a club and was going to chase him away but didn't succeed, so dad shot the bull and contacted the neighbor who owned he animal; he was pleased that the bull was dead before someone was injured, then sued the Ontario Hydro and was paid for his loss, in fact he thanked dad for shooting the bull and wanted to give him part of the meat.

While we lived on my uncle's farm and dad was working it, he had no time for holidays or even going fishing. Needless to say, we both missed that time together that we had enjoyed on the valley farm.

My dad was very independent, his parents were quite wealthy and owned a leather business in Leeds England when they died. When my dad was 4 years of age, he was left with his brothers and sisters with an uncle until he left for Canada. Well, anyway, one of dad's aunts Lady Margaret Broughton wrote to him when I was about 8 years of age and wanted him to brig his family to England to live with her until she passed on, but he decided not to accept, also my mother may not have wanted to sail on the ocean. Dad's aunt was quite wealthy and offered to leave her wealth to him if he would go back but he turned it down, as he said she should have thought of that before he left England, especially when he was a boy.

When I was 10, my parents gave up my uncle's farm and arranged to take over my grandfather Vanderburgh's farm. Mother's father was getting up in years. He was 70 and wanted to be with the family as my grandmother had passed away before mother was married. Grandfather had my aunt and uncle live with him for a few years then they decided to move back to eastern Ontario so was going to be left alone. It worked out well for both although we had to walk a couple of miles to catch a van to attend school. This was good farm land, about half way between the community centre and the valley farm, which dad continued to look after.

My grandfather Vanderburgh was born in 1851 and died in 1941, and had been a real powerful man. He had learned the carriage building and blacksmith trade, also carpentry as a young man and he had worked for seventeen cents per day for four years while an apprentice. He then opened his own shop in Delhi Ontario where he also became the town's policeman, then the notary public, insurance salesman, and also a real estate salesman. He said selling insurance in those days was like selling snowballs to Eskimos. While he was a policeman he had to occasionally attend to someone who had consumed too much beer, and on one occasion he arrested a middle aged German fellow who went up before the judge. When the judge asked the prisoner how much he drank, he said "vell I don't know, usually 60 or 70 classes but maybe I drank a few more this time. I'm sorry judge." My grandfather became quite involved in real estate and he became quite wealthy, however like a lot of people in real estate over invested and a depression hit about 1889, he lost everything and he told me one morning he woke up and didn't own the house he was sleeping in. He didn't have his shop anymore so he went to Montreal to build grain elevators which were in demand there. This he did and finally went to Port Arthur and took his son, my uncle Will with him where they continued on grain elevator construction until my uncle had a serious fall but was not killed.

They both left there and went 45 miles east of Port Arthur and homesteaded, each taking a ¼ section which a homestead was 160 acres of land, which was procured for a small fee from the government branch and had to be improved by building a house and clear so many acres and prove this was your residence for three years. They cornered each other between the Coldwater and Spring Creek, and this was about 1900 and this was where I spent my time on grandfather's farm. Some years later I owned my uncle's farm until I sold it.

Grandfather had an average frame two storey home, one storey barn for 3 horses and about 8 cows and a blacksmith and carriage shop. I spent a lot of time helping turn the forge or whatever a boy could do in this shop, as it was really interesting to me. Grandpa taught me quite a lot about carpentry, in fact enough that a few years later I took a position as an experienced carpenter, however I also know this type of work came to me quite easily, whereas my two brothers became real good mechanics. Grandpa was quite active in the farmer's association and the Liberal Party and retired from these in his 80's. He had been that was a real lesson; never complain as dad informed all of us and especially me that when you are doing your homework you don't hear anything else if your mind is properly functioning on what you are supposed to do, and eventually I found by concentrating, I could shut out all other sounds, which no doubt has been helpful over the years.

Occasionally through this period there were accidents. I recall a couple that my dad had. One time when he was using our driving mare on the buggy he had gone out to my uncle's store to pick up his sister and her family. They were Baptist missionaries on leave from China; well dad picked them up and was driving back home, there were two hills and two streams on this road, and as they broke over the top of one hill, one of the hold back straps broke on the horse's harness, this let the buggy run up close and she tried to run away. Well dad pulled her back and stopped her until they got halfway down the hill but there was a short flat space in the middle of this hill and dad had to let the mare out a little to cross the flat area, then when the mare felt this release on the reins, she jumped forward with such force that she jerked the lines from dad and threw him out of the buggy, however as he went out, his feet caught in the spokes of the right front wheel and threw him around in circles until the spokes all broke, then he was loose. The mare continued down the hill with the rest of the family in the buggy, the buggy would pitch first to the left then to the right and as they came to the bottom of the hill the buggy veered so fast it threw everyone out and they landed in a large red willow bush, the mare and buggy continued on where the buggy caught the bridge railing and skidded up to the middle of the river where it broke loose from the mare and tumbled into the river. My aunt and uncle and one cousin were only shaken up, but my youngest cousin had his leg broken, he was about 2 years old. It could have been a much more serious accident. My dad, although he had his muscles and joints pulled and stretched severely, walked down the hill from where he had been thrown out to where his sister and family were, sat down when he saw how they were, and never walked again for over 5 months. The mare came home with one buggy shaft and part of her harness, and jumped over a farm gate and went out in the pasture. This happened in the early month of May. The doctor said dad would not walk again, however dad had a bed in the living room so he could be with the rest of the family. One day along in August, when we were having lunch in the kitchen, dad hollered and almost frightened everyone, and when we went to see what had happened, he said "I moved my big toe and now I know I'm going to walk again" and about 2 months later he was walking really good. He said that ever since he had been in bed he had tried to move his toes and with determination he had done it and through his efforts was back on his feet with little affects. Dad was 46 that year.

Another time a year or so earlier he had lost part of his toes when an old hay baler slipped off some blocking onto his toes, which cut of the toes of his boot and part of his toes; he was laid up a short while but taught himself to walk without a limp as most people do limp if the big toe is missing. Dad lived to be 91 and very seldom you would see him limp. This hay baler was quite a machine; the hay was pitched into the hopper and the plunger to press the bail and was activated by a horse on a draw bar that went around in circles. One time I was driving the horse when the bar at its highest tension point came back in reverse and struck my legs, luckily they were not broken.

My second sister was born, now our family was complete. I had 2 brothers and 2 sisters. As a family we used to have some fun both winter and summer. Sometimes on a Sunday we would go on a picnic to Black Bay which was a bay on Lake Superior about 6 or 7 miles from home. Mother would make a big lunch and we would have to wash the truck and get it all shined up for the picnic. Most Sunday's my sisters and brothers and I had to go to Sunday school at the Baptist Church which was about 1 ½ miles away and occasionally mother and dad would also go.

Once in a while mother and dad would go to the city and be gone for the day, and we would be left to look after things and of course I was the oldest so was responsible for everyone. Sometime this would get a little hectic. I recall one time they were away, my brothers and I decided to do a little rodeoing; we made up a type of saddle with stirrups then got a 2 year old steer and put the saddle on him, then put my youngest brother Lorne on the steer and turned him loose, well he made a few bucks and threw Lorne off then started to run and Lorne still had one foot in the stirrup and was dragged beside the steer and as he ran we couldn't catch the steer. Luckily the stirrup strap broke without Lorne being seriously hurt, so we gave up on that as the rest of us were not brave enough to try that again. Another time when mother and dad were away, my brothers tried to make some fudge, well after they had it made no one would eat it as there was something wrong, so they couldn't leave it around where mother and dad would know about it so they gave it to one of the dogs, and this dog had it stuck in her mouth for several hours and then we thought we would be in real trouble if the dog couldn't eat or drink but it eventually melted or softened so she could get some water.

One summer we were out at the valley farm, I remember mother was there, and some times we would go there for the day, go fishing in the river and have a picnic while dad was doing some work around the farm. However this day our collie dog Major cornered a groundhog and of course we were encouraging the dog on and mother came to see what we were doing, and of course she stopped us, but while she was there she almost stepped on a baby groundhog. The dog had killed the mother so we took the baby home to feed it, as it turned out we named it Brownie and it became a real good pet and was also a good watch dog. Whenever anything strange happened or someone came to visit, Brownie would sit up and whistle real loud. He liked cake and whenever he thought he should have some, he would scamper in the house and go to the cupboard where the cake was kept and whistle until someone gave him a handout. Our neighbor across the road about 100 yards would sometimes give Brownie a handout and occasionally he would go over there and whistle at her door until she gave him something. The only problem we had was their dog would have killed him if he could catch him. Our dog Major and Brownie were real pals although Major would kill other groundhogs if he caught them. Brownie was a real clean animal and knew each of the family and would really make a fuss over anyone that may have been away for a day. He would jump up and down, roll over and sit up and whistle. We had him for quite a while. He would hibernate for the winter, however one spring he never showed up and we found him in the root cellar there he either froze or drowned as there was ice and water where he was hibernating.

One summer just at the beginning of summer holidays, dad asked me to take a hoe and my bike and go to the valley farm to hoe a patch of strawberries, and as I was peddling along with the hoe across the handlebars, I came to a little dip in the road and I would peddle real fast to help me get up the incline on the opposite side, and the last dip before the farm, the hoe caught in some brush and I went over the front of the bike and landed on my face in the gravel, scratched my face so bad I couldn't go anywhere that summer I think I was about 16 as I missed all the fun at the dances and parties right through the holidays.

One summer while haying, I had a narrow escape of what could have been a serious injury. The same mare that ran away with dad tried to run away with me while hitched to a hay rake. The rake is about 10 feet wide with curved teeth the full length, and all you have to ride on is a metal seat above the teeth, but fortunately it was near the cattle barn and root cellar and I managed to steer the mare into a corner beside the root cellar and a fence where she couldn't go any further and fell down and rolled over and got real tangled up with broken shaves, harness, and fence, and any way I only had one finger injured when the rake teeth flew up in the air and hit the seat I had one hand hanging on to, but it could have been much worse if I had fallen off in front of the rake teeth.

We had a large garden and through the experience of selling market garden produce in the summers, I found I could sell and liked it. One fall, a young lady called on my mother. She was selling subscriptions for the Curtis Publishing Company; Ladies Home Journal, Country Gentleman, and Saturday Evening Post which my mother bought, so I talked to this lady about how to do something like that as I wanted to start selling also, and I was 12 years old and as soon as the first magazine arrived, I wrote the company and asked about selling for them, and they sent me all the data and I was ready for business, however dad questioned how I was going to do my chores, homework, and sell magazines, then he tried to maybe make it a little tough by saying I couldn't sell to any of the neighbors or friends in our area. Well I had that figured out shortly; there were several towns east of where we lived and I could go there. Two of the towns were Ontario Hydro communities that housed hydro employees and their families. I had to go there via C.N.R., so with dad's approval, early Saturday morning I caught the early train and proceeded forth to become a subscription salesman. I arrived in the first town early in the morning and started to call on homes. After the 2nd or 3rd house, one lady asked me if I had the approval of the town superintendent and of course I didn't, so she told me where I had to go to see him, and off I went, as no one was going to stop me, after all I had to use my own money for train fare, and as it turned out the superintendent was a very obliging and nice fellow and he gave me a letter, also asked me to come to his home for lunch and arranged for me to stay in their employee quarters for Saturday night and my meals which I would pay for when leaving. At breakfast they had grapefruit, which I thought you ate like an orange, so one of the employees showed me the proper way. I went on my way selling subscriptions and at noon I had lunch and supper with he and his wife, very nice people, were real encouraging and pleased to see me out on my own, endeavoring to make a dollar. If I remember correctly, after paying for lodging, meals, and train fare, I had 12 dollars in my pocket; even dad was surprised that I had done so well. I made several trips back to that area and other towns and always did very well and was pleased to know I could really sell.

While a boy on the farm, my dad bought a female husky dog from Alex Budge. Alex raised these dogs as sleigh dogs; this one was part Alsatian and mostly wolf. We had her bred and got four nice husky pups, and when they were large enough, we had a nice dog team of four dogs. I used them for doing many chores, such as hauling milk to the station for shipping, getting groceries, and other errands, and also entered them in the dog racing on Lake Superior in the winter. I also hauled cement one winter for road construction. One late winter morning or should I say early spring morning, I took the dog team to get some tools for dad to a timber lot he had. I had to cross the Cold Water River on the ice which was getting near break up, and by that time I was ready to return, the snow was getting too soft for the dogs to travel on it, so we stayed in the bush all day until it froze enough to travel. When I reached the river, there was water running over the ice, the dogs were hungry and wanted to get home, but they never stopped, so I hung onto the tow line and stood on the back of the sleigh and in we went. The dogs swam part of the way across and luckily the ice hadn't opened enough for the sleigh to go through and we came out on the other side, wet but still traveling. It was about 2 miles to home, and by the time I reached there, my pants were frozen solid and I could hardly walk, but we were home, thank the Lord for looking after me.

After I finished grade 10 in our country school, I took several short courses in the Port Arthur Technical on math, blueprint reading, and bookkeeping. These helped me in years to come in the various businesses that we had through the years.

As I mentioned before about going hunting for a few days every fall with dad; we both looked forward to this break. Dad had taught my brother and I how to shoot a 22 rifle and I must say we were real good at it, and we had quite a lot of practice keeping the crows out of the grain after it was cut and stocked. Crows would clean up several hundred pounds of grain in the afternoon if you didn't chase them out. Then we also had partridge hunting in the fall. We had two 22 rifles a number of years until one broke and couldn't be repaired. When it came to hunting big game, dad had his own rifle, a 38-44 long barrel, but I had none, so he would borrow one of the neighbor's and in those days, didn't have any extra ammunition to check the rifle. I know now it could have been much better by shooting a couple of rounds. Some of our neighbors didn't look after their firearms as well as dad, and I missed more game than I ever shot until I was able to afford a rifle I was used to.

Dad was an excellent marksman and I don't remember him ever missing anything he shot at and he was exceptionally fast, and could shoot from any position. I recall one time we were walking into an area where there were mule deer and large white tail deer. I was walking behind dad, and all of a sudden he shot right from the hip so fast it really startled me, and I stepped up beside him and asked what he shot at. He said "I just shot a large buck under a spruce tree, he turned to look at us and was about to leave, so I shot him." When we got up to the deer, it was dead. Dad had shot it right in the center of its head between the eyes. Needless to say, that was the hunting for that day as we had to dress and skin it then pack the meat back home.

Dad always told me that a moose will not attack a man and not to listen to some of the stories we would occasionally hear from some of the so called hunters. However there was a day to come when he would change his mind, which I will go into later.

There was a man who lived in a cabin several miles from our home in the Ouimet Canyon area. He had a nice large cabin and asked dad and I to stay with him whenever we went hunting for a few days, so this we appreciated and did. Bill Oliver was a retired timber cruiser and other professions in the woods business, and he was a good hunter, and on occasion shot two deer with one shot. He shot a deer in some tall grass, the bullet passed through the deer and struck another one in the forehead, killing the second deer, of course he hadn't seen the second one until he walked over to the one he had shot, and there was the second one lying there also. This happened one fall when we were there, only a few yards from the cabin. One thing we did was never waste any meat that we would shoot, as hunting in those days was for the winter meat supply and not for sport, as it mostly is today. Bill Oliver had no fear of animals, and one fall that we were at his cabin, he took a shot gun out to shoot a moose that was feeding close to the cabin. We could see it feeding, and it had a large set of antlers. Bill stayed on the windward side of the moose, stayed down low, and went right up to the moose, stood straight and shot it with the shot gun, and if he had missed, it could have killed him, but one thing he had to his advantage, he had the shell loaded with several pieces of spike about ¼ inch long, which he had cut and packed himself, and I know one thing, I would never had tried this. Dad said to him afterwards why he didn't use a rifle, it would have been much safer, and he said he wanted to try the shot gun and see how good this type of shell worked and wasn't worried about it not working as he had planned. Bill seemed to have no fear of anything, was a good sized man and had a lot of experiences which made him the type of person he was. He came and lived with us and helped out with the farming while dad was laid up after the mare ran away and injured his back and legs. Bill didn't like farming but was certainly appreciated at the time.

One time we were at Oliver's cabin, there were three of us hunting in an area where the timber had been cut, and there were skidding trails where the timber had been hauled out. We would walk down these trails about a hundred yards apart and were able to get deer or moose as they would run from one trail to another. This day we had separated and traveled only about five minutes when we heard dad's rifle, four shots in quick succession, then one more. We waited for him to holler but he didn't so after a few minutes, the third fellow that was with dad and I that day, a neighbor by the name of Boyd, about dad's age, came over to me and said "I don't think that was your dad as it had to be an automatic rifle to shoot that fast," but it sounded like dad's rifle. Well Boyd said that no one can shoot that fast with a lever action rifle, any way shortly after we had the discussion dad came along and said "didn't you hear me shooting? I need some help with a moose," and in the next breath he said to me "don't let me or anyone else tell you that a moose will not attack you, this one attacked me, and that is why you heard so many shots so fast, as I kept on shooting as long as he stood up, the last shot was to finish him off and make sure he was dead." This bull had

22 points on his horns, was quite an old one and dad saw him fighting with a younger bull, and as soon as the younger bull ran away he turned and started to attack dad and that is when he shot him. The moose dropped right at dad's feet, and dad said he thought the moose didn't know he was there as he was hidden in some small evergreen trees, but the old bull must have smelled dad and knew there was something else to fight with. As it turned out, this old bull was not suitable for good edible meat. The hide on his back was over ¼ inch thick, and he had got heated up with the other bull which makes poor quality meat. We did make some of it into hamburger which was mixed with pork but most of it went for dog food and chicken food. I hauled that meat and hide and headed home with our dog team.

When I mention the dog team, I was the one who looked after them. We had 4 dogs, and our lead dog was a female ¾ wolf and ¼ Alsatian which my dad bought from a fellow who lived in the same area as Bill Oliver. He raised huskies for sale, and he would take an Alsatian female, tie them out in the bush and the timber wolf males would breed his dogs, bred and rebred back until he had full blooded timber wolves which he trained for sleigh dogs. I have seen this man, Alex Budge with 7 of these huskies in a team, beautiful very large wolves which were very docile in most cases and exceptionally good sleigh dogs. The female he sold dad was a little smaller than he wanted, so sold it for less. I don't know what dad paid, however I do know that Budge was selling the large huskies for $175.00 each in the thirties. We had a black and white border collie, a good sized dog for that breed. He bred the female husky and she had three pups. They were all different. One had short hair like the female but the same color as the collie, and this one was a little lazy as a sleigh dog and was inclined to be vicious at times. The 2nd was a very tall short gray haired wolf with piercing gray eyes, very gentle and a real good sleigh dog. The 3rd was larger again, with long light brown hair, brown eyes, and what one would say as a very large wolf, very gentle, and seemed to always have a smile on his face As pups, they were all a real roly poly group, however the mother was a very possessive and we were very careful around her while the pups were young, pups that are to be used as sleigh dogs are not made a fuss over or handled very much as they are strictly raised for work. When they were old enough to start training them for sleigh dogs, I had to do this. The female had been trained before dad bought her, so I would harness her, then take one of the pups, which was 2/3 grown, put him in a harness behind the female, hook them to a sleigh and tell her to mush, which she knew as go forward. Sometimes she had to drag the pup until he caught onto standing on his feet and following her, and it didn't take long with the two large pups, but the one who looked like his sire was real bocky, however eventually learned how but never liked it. We used him for a heel dog or next to the sleigh, also tried to bite me once and the mark is still on my face. The female became a very good lead dog; knew all the commands and eventually we had a very good dog team.

Ralph 1932

A dog can pull an average weight of 150 pounds, some much more, however, ours did much better at times, for instance that moose that attacked dad weighed well over 1000 pounds and the dogs hauled it over nine miles in one load through an area that we had to break a lot of the trail in the snow. It took a full day to go in and out with the moose through a brush trail. Sleigh dogs in most cases are kept on chains next to their dog houses, which we did with them, however the dog with the happy face, long light brown hair was sometimes allowed to run free with the collie and naturally picked up his tricks and he became a real good watch dog, as was his sire, even learned to bring people to the house who may have come for some reason or another and wandered around without one of the family. The collie would show a person to the house by growling or showing his teeth if they didn't go where he wanted. A person could go anywhere they wanted as long as long as they didn't touch anything, and if they did, to the house or to one of us if we were outside, however the big dog with the happy face would not growl or anything. He would see you put your hand down on something even to rest, then he would gently take you by the arm and lead you to the house or one of the family, and he was so large no one ever tried to go any other way than where he wanted. It became quite a joke in the family about the Baptist Church minister who quite often used to take a short cut across our farm on his way to church as he usually walked, and he would occasionally stop in the barn yard and rest while going through, and one of the dogs and sometimes both of them would bring the minister to the house before he could go on his way.

This dog team was real handy around the farm in the winter. We hauled a lot of things, including the milk that had to be taken 6 miles to the C.P.R. station to ship to the dairy; also hauled dynamite one winter for the construction of one of the roads in our area. We used to enter the dog derby on the 14th of March each year on Black Bay, a part of Lake Superior bordering our community. Never 1st, but was always 2nd or 3rd place and was pleased to be there as there were some real good dog teams in our area during the thirties.

Back to school days, and while attending school up to grade 7 or 8, I had a teacher from Manatoulin Island, Miss Peggy Robertson who later married and was Mrs. Peggy Bohler. She was a real encouragement to any pupil that listened, always told us you can be anyone you want to be, set your goals and keep your head up. She often told me I should be an author as I had such an imagination, and liked to write short stories which encouraged me and would read them to the class whenever there was time. One time I won a book in a contest for writing. I can say that all the way through my school, I always had good teachers, of course I enjoyed going to school and wanted to go to university, however due to the depression of the 1930's, I didn't make the grade, but as my grandpa would tell me "make up your mind you can do whatever you desire with understanding and determination."

After finishing grade ten, I did take several short courses in the Port Arthur Tec as previously mentioned and always read anything that would improve my ability and understanding. I feel a person never gets too old to learn something.

After we moved to my grandfather's farm, we had to attend

church. It was a community Baptist Church about a mile from home. I remember giving my life to the Lord when I was about 12 years. I also strayed a long way from Christianity for quite a few years but realized that through Christ all things can be accomplished and thanks to our daughter in later years did get the family back in the church again. There is no doubt in my mind that Sunday school I attended helped me to try to be honest in my dealings and business throughout my life.

Prior to the hungry thirties, for several years dad did some pulpwood contracting and also farmed, so we had a bunkhouse for these men that worked in the bush; they also ate their meals in our home with the rest of the family and these men were from European extractions and most of them could talk very little English if any. Dad would tell me what he wanted and I was supposed to make them understand, so I could get by and partially speak their language with some English. I recall telling a fellow how to harness a team of horses. When we went to see how this fellow was making out, he had the harness on backwards and didn't know what the horse collars were for, so that was one time I really didn't make the fellow understand, so dad showed him how the harness went on. He thought the britchen was the breast collar, and the hames were ornaments. Some of the fellows became good friends and worked for dad for several years, and some of them he would keep through the summer to work on the farm as the brush operation were just through the winter.

We used to have a lot of fun during the winters with toboggan slides or skiing. We would gather at some prearranged home where there was a hill suitable for toboggan sliding. We would slide for 2 to 4 hours then go in the house for cake and hot drinks. One time we were sliding and we had some long toboggans where up to 6 or 7 teenagers could ride together, so my neighbor Norman who was a good sized teenager, in this case was going to steer, and there were several heavy girls sitting on behind him, so the more weight on the toboggan, the faster you would travel, and as it turned out, they may have had too much speed as part way down the hill Norman lost control and they went over the bank, went down quite a distance and hit a tree or stump, Norman slid forward off the toboggan and lost the seat of his pants. He was so embarrassed that he went home and didn't stay for cake and drinks that time. There were some real good hills around the area, some hills we considered too steep and fast for the average toboggan. My brother and

I cleared a hill on our farm that was real steep; it was in a good area with no side banks. We wanted to go further on the flat at the bottom of the hill, so besides being steep, we iced the hill to gain more speed, it worked too well and Wray was nearly killed on it as he landed on his head at the bottom of the hill and was fortunate that he didn't break his neck; none of the regular group wanted to use it, so we went back to slower hills.

As to skiing, I had a set that were homemade and not the best, in fact very crude, but I learned to ski with them. Eventually I bought a nice pair from a Swede fellow who had made them, and they were really nice skis. There were several Swede men who worked for the local saw mill and they would go cross country or mountain trails and after I bought the skis, they wanted me to go with them which was usually on moonlight nights, and I became quite good and really enjoyed skiing. One time I had an experience that really frightened me. We were skiing in a mountainous area and there were only trails cut through the bush, and this time everyone was ahead of me, and when I came off a steep run, I didn't make the turn and landed headfirst in a large red willow bush which had 3 or 4 feet of snow on top of it, and I broke right through the snow, still had my skis on, and I couldn't touch the ground as I went in head first with my hands. Every time I moved, the snow would pile in around my head. I was afraid I was going to suffocate before I could get out; however the other fellows came back looking for me when I hadn't arrived at the bottom of the run. They helped me out, but I never wanted to have that happen to me again; but it didn't stop our skiing, and was more cautious not to land on another willow bush if I fell.

When I was going to public school there was always some bullies as normal and also those who thought they were quite good at fisticuffs, as it turned out as I became a teenager I was quite well developed from working on the farm, and as for fighting that was a no no, for when I came home, I would have more trouble than fighting back at school, so due to this, a number of boys would pick on me, trying to make me fight, even younger and smaller than I was. One time I came home and told dad about this and we had a good talk out in the barn about fighting, however he did say "don't let these boys pick on you, and if you have to fight, don't talk about it in front of your mother."

Dad said "we will have a real sissy on our hands." Mother didn't want any fighting but I knew I had dad on my side and mother was quite easy to talk to, so I knew what I was going to do. There were some boys that I was going to even up the score with. The next time someone started to bully me they soon found out that I could look after myself, but this can go the other way also, then I became the target as to who was the best fighter in school, and I found there was the occasional fellow who was my match, of course this helps a person to smarten up and not always have a chip on one's shoulder, and also try to avoid fighting if possible.

When growing up, we had two 22 rifles and dad taught my brothers and I how to shoot, and we all became very good marksmen with practice. We had to keep the crows and rodents out of the grain fields and therefore we did quite a lot of practice. There was also partridge hunting in the fall so that my brother Wray and I were using the two rifles whenever there was an occasion. One time dad was using the older 22 rifle and it blew apart, unfortunately he was not injured, but we only had one rifle now, and that was not good as every time I wanted to use the rifle so did Wray, and there used to be some arguments about who was first to use the firearm. It wasn't long before grandpa made the decision. One day when we both wanted the rifle to go hunting partridge, grandpa said "I'm tired of this arguing, I'll fix this," so took his old Snyder long barrel single shotgun out and said to me "you are the oldest, come with me and I'll show you how to use this gun," I refused but that made no difference, so he put a large piece of cardboard on a tree and proceeded to show me how to use that old Snyder shooting iron. It held one shell which he put in, and you could use a single shot shell or buckshot shell in this gun, but this time it was buckshot. He was about 50 yards from the cardboard, put the gun to his shoulder, and said to pull it in tight, hold, then pull the trigger. "Boom," my grandfather was a strong well built man, and when I saw the kick that old firearm had, as grandpa's shoulder really moved back, I said "no way am I using that." I think I was about 12 years of age at the time, however no matter what I said, it was decided then and there that I was going to use the "blunderbuss" if I wanted to go hunting, so there too. Something I should say was, when the shot hit the cardboard it disintegrated. Well anyway, one day I wanted to go partridge hunting after I figured everything had cooled down, and tried to get out with the 22 rifle, but grandfather caught me and said to use the Snyder, get use to it and you will like it, so there I was, no hunting or the Snyder, so I took the old "blunderbuss" for a walk, all the time thinking I'll never shoot anything with it, as I could see grandpa's shoulder kick back 2 or 3 inches when he pulled the trigger. I think I was hoping I wouldn't see anything to shoot at, any way I was walking along a side hill along Spring Creek when I saw a partridge sitting on a knoll, and as I had been taught to shoot fast when you saw a partridge or any other game, I pulled the old Snyder and fired before I thought about what would happen and the old boomer kicked me so hard I went over backwards. When I got up, the knoll the partridge was sitting on looked like a rotovated field, and the partridge was sitting in a tree above the knoll, looking down chattering away like a chipmunk. I left that partridge there as I felt if it was the same one, it deserved to live, and anyway I always had a lot of respect for that old Snyder, and eventually liked using it except for the weight. The family used to kid me later on that I would find a bunch of partridge, bunch them all up with the long barrel then back off a way and shoot the whole bunch with one shot; however I had shot 2 or 3 birds with one shot as the shot spread was quite wide at 60 to 75 yards up to a 3 feet spread.

Dad had told me he had shot a moose with one slug in that Snyder at over a quarter mile, so he said you had better respect that old gun. The worse part of that whole experience was my brother Wray has it for an antique and I was the one that had to use it.

I'm reminded of a story about a Swede fellow who had a vehicle in the early 30's. Ollie was used to getting up to a cup of coffee prepared by his wife Olga. Well, as it turned out one morning Olga didn't have the coffee ready when Ollie arose so he got a little peeved, put his cap on crossways and stormed out of the house to go to the coffee shop, went out to the garage and jumped in his truck and proceeded to back out and he hit the door jam with the fender, this made Ollie a little more excited and he forgot to stop and backed into the neighbors hedge, knocked part of it down and killed their cat which was sleeping under the hedge. Well then Ollie really got excited and drove forward and ran into another neighbor's tree and wrecked his vehicle. When someone called the police, Ollie explained that it was all Olga's fault because she didn't have his java ready when he got up and that made him come out of the house with his cap on crossways, and I'm sure you can hear Ollie's accent when he got excited explaining to the policeman.

When I was 15 years of age, dad said I could have a new bicycle or learn to drive the truck, well, I looked at it this way, I couldn't drive until I was 16 and dad would have to go wherever I went, so I bought a new C.C.M. bicycle with my own money I had earned from my row of strawberries. When we were helping on the farm dad gave Wray and I each a row of strawberries, so all the money from this row was my own for helping out on the market gardening and I remember the first year I had $24.00 in my bank, I was wealthy.

Back to learning to drive our truck, that took place when I was 16 years of age and wasn't long until I was running errands for dad as he didn't like driving, also I was going to the odd party or dance with the truck, was real proud of myself. One day I was on some errand for dad and took a little detour to see a chum of mine, was longer than I should have been and was rushing back, turned a corner too fast and ran into one of our neighbor's fence posts, so I backed up, looked to see if the truck was marked , and it wasn't, fixed the fence as fast as I could and headed home and didn't say anything about this incident to anyone, however a few days later the neighbor stopped in and he was asked to stop for lunch. While we were eating, he asked if our truck had run into his fence, and dad said not to his knowledge, of course he didn't know about it and no one disputed dad's word, so the neighbor said it looked a lot like our truck pulling away from his fence corner, however, fortunately for me there was some brush there that helped to obscure the view, and I wasn't about to say anything as I would have been grounded for a while and didn't like riding the bike any more than dad liked driving.

When I was nearing school completion, grade 10, I talked to dad about going to college, this was 1933, and dad said he would like to see me go, however we didn't have the funds as we were into the hungry thirties as that time was known, but he said "I think I can get you a government position and you can save your money and go back to school later on.

Ralph

The winter of 1933, 1934 I worked for a short while for the Provincial Government at a trout hatchery helping to put up ice which they needed during the summer for distribution of the fingerling trout that were distributed from this hatchery, then in April of '34, I was called in to be in charge of the distribution of the young fish at very good wages for those years, $65.00 per month and all expenses paid when I was out working away from the hatchery. I liked the work, was traveling all over northwestern Ontario, about 2 to 3 hundred miles east and west of the hatchery which was at Dorion, the same community I grew up in, which is about 45 miles east of Thunder Bay (formerly Port Arthur.) I like the bush, lakes, and rivers, traveled by truck, railroad, and air, and also canoe in some remote areas, and a lot of walking. These fingerling trout 2 to 4 inches long, carried in tanks of water weighed about eighty pounds and could be packed on your back, of which I carried many to stock various stream, rivers, and lakes. The ice we had put up the previous winter was to keep the fingerlings cool between the hatchery and point of stocking.

When traveling by rail, which would be in the baggage car, I had to shake the containers quite regularly to keep the oxygen in the water to keep the fingerlings alive, and then you also placed ice on the top of the container which helped to give off oxygen. One thing I did find, that when traveling on the C.P.R., it was much rougher and required less work than traveling on the C.N.R. which has a much smoother roadbed. While working at this position which normally ran from May to October, I was on call 24 hours per day and therefore worked all types of hours.

Sometimes during the summer I would get real tired. I remember one time coming home one day before noon to have some rest, went to bed and was dreaming I had to catch a certain train or something, got out of bed and was heading out the door when my mother caught me as I had my clothes on my arm and was in my sleepwear, this time I was traveling in my sleep. She woke me up and I went back to bed. I was so used to getting up at any hour that usually I could go to bed at any hour I wanted to. One time I took a room at a hotel in Armstrong in northern Ontario on the C.N.R. line. I left a call for 2 am as I hadn't had sleep for quite a while and had to catch a train at 4:30 am, however as usual I woke up before 4 am and when the hotel owner came to wake me and found I was awake, he almost chased me out of the hotel, so I guess I really disturbed his sleep with this call.

I had some real experiences while distributing these fingerlings, as I said by all types of transportation, also packed these small trout into many lakes , would meet a group who wanted a lake stocked, and off we would go. I packed many tanks of fingerlings over the 3 or 4 years. One time I was with a fellow in a canoe who got so excited when he got in some rough water that he started running down the canoe which is a no no, so I yelled at him to sit down or I would hit him with the paddle as I didn't' plan on drowning in that area, so he realized what he was doing and calmly sat down and I got us in to shore as quick as possible. When we landed, he apologized for his behaviour and he had done lots of canoeing and didn't know why he panicked but had never been in rough water like that before which was caused by wind in the narrows between two lakes.

In the late summer of 1937, I had finished distribution of fingerlings at the trout hatchery, so decided to look around for other work, as the bush work had not started up yet. I went to a company at Red Rock where a pulp mill and town was being built, and required the position of a town clerk; this position entailed records of building materials received and record of where used, keep employees time records, and generally answering to the town superintendent. It meant I was really busy, especially in time, when the superintendent began drinking considerably and more work was piled on me. Eventually there was changes made of construction engineers but the work was the same, however due to financial difficulty by the investors and one of the investors being asked to leave, the project closed. I had worked about

3 ½ months. The investor that was asked to leave was R.O. Sweezey of Montreal who financed many projects, and he eventually bought out the other investor's shares at a reduced price then completed the mill and town about two years later.

When I was employed at the pulp mill construction, I purchased my first car, a small down payment and monthly payments. I bought this vehicle so I could travel on weekends to see my parents at Dorion, but there were several fellows who wanted transportation to Port Arthur on the weekends also, and with their money I could make my monthly payments on this car which was a 1938 Pontiac sedan with room for five passengers and the driver. I remember it was a very heavy vehicle with a flat head eight cylinder engine and could really roll, as it turned out, in more ways than anticipated.

About halfway through the project we were in the city of Port Arthur for the weekend and had my paying passengers with me. Monday morning we left the city early to get back to work which was about fifty miles. As I had gone to the midnight show at the theatre with my girlfriend and had not had much sleep, I asked a friend of mine to drive as he was one of the paying passengers. About 20 miles from Port Arthur we had a flat tire and he was driving quite fast, he pulled on the emergency brake thinking he would save the tire, well he really did it. When the brakes were applied the wheels stopped but the car took a somersault, then slid on the roof, then turned sideways and started rolling. As there was only a small area of the roof that was metal, the rest of the roof peeled off and when the car started to roll, it was flinging passengers out as it rolled. Needless to say it was the end of the car but luckily no one was hurt seriously, but my girlfriend was bruised up quite a lot who was going with us part way to work, and the driver was pinned in the driver's seat, the only metal on the roof fit down on his head like a cap, and he couldn't move without our help. This was the end of my first car, and the end of the payments, and so was out of business so to speak, luckily I had insurance but did not buy another car for some time.

As it was getting time to go back in the pulp camp for the winter, that was my intention, however dad had other plans and wanted me to operate a milk pasteurizing plant at the small town of Cameron Falls he had acquired through non payment for milk shipped to this small dairy over a period of time. He suggested I run the dairy and distribute the milk, receive my board and room, and my wages would go to buying half interest in the business, this of course looked interesting and I was always ready for adventure, so off I went to operate a dairy, a few instructions on operating the pasteurizing equipment and I was in the milk retail business. The milk arrived at the dairy by rail in 8 gallon cans daily, and then I ran it through the pasteurizer, and then delivered the milk to our customers.

This was the same town that I started selling subscriptions for the Curtis Publishing Company when I was about 12 years of age.

Ralph – Owner & Operator of Dairy

After about 2 months we moved this dairy to a mining town on the main line of the C.N.R. in Beardmore. We had to construct a building for the pasteurizing machine, a boiler house for steam heat and hot water, also a cooler for the milk, and a small house to live in, which were 2 rooms. It only took a few days to move the dairy to the new location; now to start calling on every home and business in the area, the population was about 3000 at that time. We bought a delivery wagon and a real good horse for hauling the milk. It was shipped from dairy farms at Pass Lake and Dorion by rail daily. I picked it up with the horse and delivery wagon, took it to the dairy where I pasteurized and bottled the milk, then delivered it to the customers. This usually took about 8 hours per day from 6 am to 2 pm, then any milk that wasn't sold would be used for making cottage cheese which usually took about 3 hours per 30 to 50 pounds, not profitable but we did get some income from the extra milk which we had to pay the farmers for, always tried to keep the supply as close to retail as possible but it was difficult as our wholesale customers would change quantities according to the meals served in a restaurant for instance, however, although I never earned any more than my board for about a years work it was a real educational experience. My brother Wray helped for a few months. This business came to a sudden end one morning when the dairy caught on fire while getting up a head of steam. We think some fire got into the shavings which we used for firing the boiler while we were away from the building having breakfast. The buildings were insured by a company who stalled on payment for over one year and 3 days, so if a claim is not settled or suit filed against the insurance company before that period, they can settle any way they wish. We of course didn't know this until it was too late and they returned the fee we had paid for the insurance. Not having the capitol to start again which was $7000.00, the business ended there.

The spring of 1934 was warm and dry in Thunder Bay and surrounding area. I started getting itchy feet and wanted to try to get work somewhere other than the farm, and the trout hatchery didn't need me for a month or two. The area northeast of Nipigon along the C.N.R. was producing some mines and there was lots of work and of course lots of men looking for work, anyway I decided I would go and try to find some work in the mining area and also see an area and things that were different. I arrived in Jellico via C.N.R. with a few dollars in my pocket but found that everything cost more money than I had anticipated so I had to find work fast, I believe I had a return ticket just in case there wasn't work for me.

The town consisted of two hotels, two boarding houses, two cafes, bakery, and a population coming and going of probably 1500. There was lots of activity and work but as I said lots of men looking for work, probably 10 for every job. One day I was in the bakery and the owner was a little French Canadian said he needed a chore boy, so now I had work although not much wages, however it would keep me going until something better was available. I worked there for two weeks then went out for the forestry fighting a forest fire around the Black Water area. Something I thought was amazing, Frenchy, the owner of the bakery was about 5 feet 4 inches, probably 120 pounds soaking wet and used to show me how strong he was by picking up a hundred pound bag of flour in each hand and would actually put it above his head. He thought he was tough and as there was usually a fist fight or two around the hotel every night, he would go there to try to get into the activity. One evening two or three fellows tried to pick a fight with Eli Cook who ran a freighting business along the river for the prospectors and mining company. Eli Cook kept telling these fellows to buzz off and leave him alone; Eli was a big man and also very strong. These fellows kept bothering him and taking the odd poke at him until he reached down and caught a man in each hand by the pant leg, lifted them up off the ground and gave them each a shake and as they fell back to the ground Eli had a piece of cloth in each hand from their pants then turned and walk away. I think after that, Frenchy quit bragging about his fighting ability.

    As I said before, I went out for the forestry department fighting fires, this lasted about six weeks but was very interesting and at times awesome. There had been a very dry hot spring not normal for northwestern Ontario. This fire apparently was started west of Port Arthur at a camp and with the high winds, traveled east and north to the Black River area. There are many lakes and rivers in the area and at that time the airplane was not used for fighting fires. We used canoes for transporting men and supplies, and as I was well accustomed to being in the bush alone, the chief fire warden sent me out scouting for locations to set up pumping equipment ahead of the fire; this sometimes was difficult as the wind would start blowing real strong and the fire would pass me, so as soon as you saw that happening you sunk the canoe and would lay in the water with only your nose out to keep from being burnt, and it happened to me several times, in fact when I first started, I had a young Indian with me, and after several fast dips because of the fire, he wouldn't travel with me any more. We did lose fire equipment several times because of the winds but no men, although I did see a deer burned caused by the changing winds, or the animals could not get away fast enough. One morning while walking along the shore of Expansion Lake, I saw the largest tamarack tree I had ever seen, it was about 2 ½ feet through which was quite large for trees in that area, and quite a few years previous to 1934 a disease had gone through northwestern Ontario which killed all tamarack and there were very few large trees left. Most of the tamarack was less than 3 inches at the butt. Going back to the fire and the speed the wind would carry it, this became very serious for two prospectors, and one day they had been caught on a back fire and paddled their canoe out to an island where they would be safe, however the island caught on fire and instead of getting in the water with their canoe, they turned it upside down and crawled under it, the canoe caught on fire, one fellow lost his eyesight, but I'm not sure if it was permanent, and the other was burnt quite badly on his back and had some real deep scars. There is a story to this happening. These fellows were taken to a hospital by a plane that picked them up at the site; however about 35 years later one of these fellows, Jack Stinson was having coffee at our café near Princeton B.C. and we were discussing past experiences, and happened to mention the fire of 1934 near Jellico and about these two fellows under a canoe that burnt. Jack jumped up and pulled his shirt off to show us his back which still had the scars. At the time of this incident I didn't know his name or much about him other than they had been prospecting in the area.

Another thing that I remember about something that happened during the fire, we were running short of food so I had a fish line which I usually carried when in the bush, as there are fish in all the lakes in that area. The fire chief asked a couple of us to try to get some fish to supplement our food, which I believe the cook said we were down to dried potatoes, canned milk, canned tomatoes, and tea. We caught a few pike each morning. Bill Elliot and I would go out and troll the line for our daily meals, and one particular morning we were out and a young fellow from Sudbury wanted to go with us so we agreed, but he didn't know how to paddle a canoe so put him in the center. This was an 18 foot freighter canoe, I was in the bow and Bill was in the stern. Sudbury had the line, and very shortly he had a large pike on and asked what he should do now, it was really jerking on the line so we said to wind it in, which you had to do as the line was not a rod but a short block of wood which we wound the line around as we pulled the fish in, this time we had quite a large pike and when Sudbury had it near the stern of the canoe, Bill who was always joking about something said "boy have we a large one this time, I think it's larger than the canoe," and Sudbury said "what do I do now?" and Bill jokingly told him to throw the line overboard or it will sink the canoe, and what do you think? Sudbury threw the line in the water and it immediately disappeared with me hollering to catch that line, needless to say, that was the end of our fish meals, so back to dried potatoes and tea.

About that time of day or later, the weather changed and it rained then snowed and that was the end of fire fighting in this area. We started packing and loading equipment to move out to Jellico. We were camped on the north side of Expansion Lake, which was as I think I said before about 4 miles wide at the narrowest point from shore with some deep bays. The morning we were ready to leave there was a light wind and we thought that this would be a good time to move as there were quite a few canoes with equipment and men. The canoe I was in had five men plus the stern man who ran the outboard motor. We came out of one of the bays into the main lake and there was a strong wind blowing and the waves were about 3 feet; in those days there were few life jackets, so of course there was some concern about staying afloat, but between bailing and a good stern man operating the motor we made it safely. We arrived in Jellico mostly dirty and wanting a warm bath, shower, or anything, as except for a dip in a lake or a river to save yourself from being burnt. Most of us hadn't had a good cleaning for a month or more. We discovered a steam bathhouse and of course everyone that could, took their turns, this was large enough to hold several people, so there were about six of us in the steam bath at a time, and of coarse someone had to start a little hellery so they poured all the water on the hot rocks to see who could stay in the steam bath the longest and how much steam each could stand, and some of the fellows were lying on the floor trying to stay the longest, however one after the other had to leave, but one tall lanky fellow rushed out and jumped in a 45 gallon barrel of stagnate water to cool off. I guess that water had been there for years, because when he riled that water we all had to scatter because of the odor, so he had to go back in the steam bath, he later told me that he could smell the rotten odor on his body after several days and baths.

The water had been there for fire protection, in fact there were 3 or 4 barrels with the rotting water and slime.

I arrived back home in Dorion about the end of May and went back to work for the Dorion trout hatchery, this time distributing the fingerlings or small trout to the various lakes and rivers around the Lakehead area. This work I did for several summers.

In the late thirties we used bush planes quite a lot as it cut down time and labor costs and could get into remote areas also. Usually we would load the plane with the containers and I would go along and dump the fingerlings. When the plane landed, up to this time it was always thought that the water temperature had to be evened out to the best of your ability while dumping these. Long about 1937, we tried dumping the fingerlings as the plane flew over the water, rigged up a harness fastened to the inside of the plane with the door off or open, and you reached for a container and poured the water and fingerlings out and after dumping your load , then fly low to see if all survived, and if any died or were killed from this method of planting fish, you could see the white bellies of them, as it turned out there were very few, and it wasn't many years later that planes were used for planting young fish that had a shoot in the underbelly of the plane and they dumped from a tank as the plane glided as low as possible over the water. I worked at this during the summer of 1934 to 1939 and enjoyed the work.

Dad was selling hay, potatoes, some vegetables, and pork to the Provincial Paper Company who owned a mill at Port Arthur and timber cutting rights a few miles from our farm, later years this became part of the Abitibi Pulp and Paper Corporation. The superintendent of the bush operation used to call around to see dad about the purchases, and one day after I had finished with the trout hatchery for the season, I was at home when the woods boss showed up, so I asked about winter work and so it was arranged that I come to camp and he would find something for me to do. Shortly after that, I arrived at the camp, they had good accommodations. I was introduced to the work foreman and assigned a bed in the bunkhouse. The first work I was assigned was what is known as road monkey. I went with an axe and bucksaw and checked all the strips as they are called "roads" and cut off any stumps that were too high for the sleigh bunks to pass over, also any trees that may have blown down, and I did this for about 2 months, then went on to another job. For this I was paid $2.00 per day plus my room and board, this was quite good pay for this type of work in those days. I worked for a short while for a dollar per day for the Dorion municipality repairing roads and no board one fall while waiting to go back to the bush. The Provincial Paper always paid top wages but they never kept anyone who was lazy or as it was said in those days, "you had to cut the mustard" to stay with the company. A year or two later when hauling pulp, I was caught on one of the high stumps and had to unload in order to move.

I worked for the company several winters, road monkey for a short while, then tractor operator's helper. We broke strips, which mean running the tractor up each road that the pulp would be hauled out by horse. We iced the main haul roads where the pulp was hauled to the landing which would be on a lake or a river, and we hauled freight, which included groceries for the men and also feed for the horses and some diesel fuel for the tractor; even hauled hay and vegetables from our farm but that was by truck. I also hauled pulp 2 or 3 winters with a team of horses. I was the teamster, as I used horses on the farm, and this time I had a helper. We were paid so much for each cord of wood delivered to the landing, and I believe I received 25c more than my helper per cord because I had to look after the horses. We hauled up to 5 cords of pulp to a load; one cord weighed approximately 4800 pounds. We usually had very nice horses to work with, some of them heavy. One spring I bought a team the company was going to retire for $175.00 which was almost my whole winter's pay, and I took them to the farm where they were used for quite a few years. This pair of horses weighing 2200 pounds each were Percherons.

Provincial Paper Camp

The team was well matched and was real nice to work with. One fall dad wanted me to do some ploughing down on my uncle's farm, so I was using a walking plough, single furrow and it was play to these big horse, they would walk almost 5 miles per hour pulling that plough. We were going along fine, and one day the plough hit a large rock that no one knew was in the field, well the plough stopped but the horses and I didn't, one of the draw pins broke and they dragged me right over the handle bars of the plough. When you work with this type of plough, you tie the reins behind your back so you can hold both handles of the plough, and until I holler whoa, these horses didn't know anything was wrong, I can say I had a stomach ache from going over the handle bars.

Back to the bush camp I cut pulp, which is quite heavy work. My brother Wray and I had gone to the camp together that fall earlier than usual but there was nothing to do except cut pulp wood. As Wray was quite light for this type of work, we decided to cut together and he would be my helper.

On the timber limits that are allotted to the various pulp and paper companies, they agree to cut all the timber or the allotted species, and in the case of Provincial Paper Company, they had to cut all evergreen trees above 4 ½ inches I believe, this means that when the cutting season started in October, each cutter was allotted a strip normally 30 yards wide. In the poorest timber and as the season progressed, which was normally Christmas or the end of December, you would be allotted better timber, this way it kept the cutter happy and because a cutter was paid by the cord cut and piled, which when I worked at this was $2.00 per cord and I paid $1.00 per day for room and board. In the Provincial Paper camps, no one could work on Sunday except for the necessary chores or repairs. I was a good cutter and I remember we had no chainsaws in those days so it was all cut with a bucksaw or swede saw with a 4 foot blade and an axe, and there was real knack to keeping your saw and axe sharp. The cutter also had to clear a road through the center of his strip, stumps no higher than about 4 inches and all brush off the strip road. You would pile the pulp lengthways in piles 4 feet high along the strip road and pulp logs had to be 8 feet 2 inches long, the extra 2 inches was to allow for damage on the end of each pulp log. When the pulp was on the drive down the rivers and lakes it was taken to the mill for processing to paper. Anyway, getting back to my cutting. I averaged 3 cords a day and at the end of the cutting season, the foreman told me there was only one man that averaged more than me, he was much stronger than I, however I was real pleased with myself for that three months. I had averaged $6.00 per day, less my board, so that was really good pay in the mid thirties, even if the work was real hard.

As I said earlier, one fall my brother Wray and I were allotted a strip of timber and we were going to work together, but it was really poor timber for pulp wood, large spruce trees with limbs right down to the ground and to make matters worse, there were large birch trees growing between the spruce. Once you had the limbs cut off the base of the tree for enough room to cut it down, you had to be careful that it didn't fall into a birch tree. We were not cutting birch, but made a lot of work if a spruce tree fell into a birch as all evergreen trees had to be cut up into pulpwood in 8 foot lengths and piled. We worked together for about 2 weeks and were not making any wages or not enough for two anyway, so the foreman gave Wray another type of work and I continued on my own.

While we were working together, we saw one of the funniest things happen that I think anyone ever saw in the pulp operation; the foreman started another fellow in the next strip to us. He had just come out to Canada from Europe, I believe Austria, and he didn't know anything about cutting or even a saw or an axe. The foreman came over to us and asked us if we would keep our eye on him and if possible give him a few pointers, he also asked the fellows on the opposite side of him to do the same. This fellows name was Joe, well he had lots of problems, and it was bad enough for an experienced woodsman to cut in this type of mixed trees, and with no experience, he never knew which way the tree was going to fall and had more evergreens fall into birch trees than all the rest of us put together. We each took turns trying to help Joe, even showing him how to sharpen his axe and saw, especially the saw. It eventually got around to letting him get along on his own as we were spending more time trying to teach him and help him than doing our own work.

Well one day shortly after deciding to let Joe go on his own, he fell a big spruce tree into a tall birch tree, we saw it happen but none of us let on we saw it, so Joe decided he would get this one down himself, he was quite a supple fellow, took his axe and walked up the sloping spruce which had pushed the birch over in an arch to about 20 feet from the ground. When he got up to where the two trees were together, Joe decided to cut the spruce off at this point of contact and ride it down, but when the spruce fell, however it went down so fast Joe grabbed onto the birch, dropped his axe, luckily he had a good hold on the birch, as it really whipped up and over the other way. We heard Joe holler whooee as he was flying through the air as the birch tree whipped back and forth several times with Joe hanging on for all his worth still hollering whooee, but if he had let go of that tree on the first or second whip, he would still be flying. With him hollering and flying back and forth, it was the funniest thing we ever saw. When the tree stopped whipping, Joe slid down to the ground, picked up his tools and headed back to camp, and I don't know what he did after that but he wasn't going to cut anymore pulpwood He was fortunate he wasn't killed at that.

One winter after Christmas, all or most of the local fellows would go home for Christmas, and some of the cutters would go out to town for a good drink, then would come back with a big head and work their hangover off. The pulp was cut as I've said before, October to the end of December then hauled out of the bush from January to the end of March. One winter the foreman gave me a team of horses that had a load of pulp dumped on them with a previous teamster and they were frightened and would balk, wouldn't work, so he said "we will pay you so much a day until you could get the horses working again." Eventually I did get them working and went on piece work and was allotted a helper, which wasn't the best, he moved too slow. When two men are loading pulp, a man stands at each end of the pile and you pick one stick at a time and place it on the sleigh, which has bunks and stakes to hold the wood on the bunks, you deck it up with the two front stakes off until you have it almost loaded, then put in your stakes and drop in 2 or 3 sticks of pulp in to hold your load. With this helper I had, quite often he would be behind me when throwing the log on the sleigh ad it would land crossways. I had complained to the foreman about this and wanted a better helper, however that didn't happen until we had an accident and I was injured, this day we were standing in deep snow loading, when we just about had the front bunks loaded and ready to put our stakes in, my helper threw a stick slow and it landed crossways and the load let loose and rolled off the bunks but I couldn't get out of the way fast enough and the pulp landed on top of me, fortunately there was enough snow to stop from crushing my bones under the bunk by the sleigh runner, and I could hear conversation going on but couldn't move, but someone say "lets get the pulp off him and see how he is," and the foreman said "no need to hurry he will be dead under that much wood." Well I hollered to move that wood fast and get me out of here. I know my back was injured but not how bad, so they got busy and were throwing the wood off me, and still hollering and I remember saying "when I get out of here I'm going to kill that helper for always being too lazy to throw the pulp straight," and the foreman said to him "you better head to camp and pick up your wages and leave as I think he means it if he can move." When I got out my helper was long gone, which was good as I'm sure what I would have done As it turned out I was in bed for several days, and the supervisor was real concerned about me, he would come in the bunkhouse every morning to see how I was. I could have gone to the hospital on compensation, however in those days; anyone who went on compensation didn't get a chance to work for the company again very often. The result of this accident was to have some changes in my life in later years.

As I've said before, I did a lot of different work for this company. I learned to scale pulp and eventually got a scaler's license, also did cruising for this company and other companies, this means calculating the amount of timber by walking through the timbered area. Today a lot of cruising is done by helicopter and photography equipment then counting the trees in those photos and in cruising you do not do every area of ground, you pick an average plot and calculate this and average it out over the total area, usually works out quite accurate.

Scaling timber is counting the number of cords, also checking the length of the pulp cut, then there is log scaling, measuring the length of each log and also butt end and top end. With these figures you can figure the number of board feet in each log. Occasionally in later years I would be hired by the forestry department to do some check scaling, which means going into a certain area and check certain bush camps to see that the regular scaler was giving the accurate scale, a check scaler was not too popular in many camps, occasionally a scaler would be paid off not to scale all the timber cut, or the company would keep the scaler drunk, and only a check scaler could find this problem, and this did happen to me in one camp of the Johnson Timber Company, they were well known to steal timber or not pay the stumpage dues. In this case the regular scaler had been drunk for several months with liquor supplied by the company. When I went into the camp and asked the camp clerk for the books pertaining to the scale of timber , he wouldn't give these to me, so I phoned our head office in Toronto and shortly after this a chief inspector arrived and had the clerk under house arrest and seized the books, and while I was waiting for the chief inspector, I was busy calculating the timber cut, and this is no easy task. I measured off an average area and measured the tops of each stump in this area, then calculated the amount of timber by measuring the total area cut over, also calculated the amount of timber dumped at their landing, and also discovered they had been cutting 9 feet 2 inches, one foot longer pulp logs instead of the regular 8 feet or 8 feet 2 inches as they were supposed to have cut. I turned in my report (estimated one million cords not scaled.) The government inspector insisted on a log count in the spring when they drove the pulp down the river. What takes place is a gate is set up at the mouth of the river and every log is counted, the timber company had to pay for this expense also pay stumpage on over a million cubic feet of timber which they were trying to steal and then they lost their timber license and had to buy timber from other logging companies in the area to keep their mill operating at Fort William. The company was real good to their bush camp clerk, as he was promoted to one of the top clerical positions at their mill in Fort William and some years later made it his business to acquire two box cars loads of lumber from a mill I had operating for me, and these two cars were never paid for due to this same clerk. I believe he did, trying to get at me for catching the company he worked for while stealing timber

    The chief inspector for the forestry department tried to get me to work permanently for the forestry and said he would recommend me for a good position as I had calculated the timber stolen very close to their count, and said I had done an excellent job. Well I strayed a little from working for the Provincial Paper Company bush work. One winter I clerked in the van which is actually their store in a bush camp, look after the sale of tobacco, chocolate bars, and etc. that the employees always bought and also tools, then there was keeping men's time, scaling the pulp cut by each man, keeping track of wood hauled, and ordering supplies, it was quite interesting. After that winter, the spring of 1940, the woods superintendent and one of the head staff for the paper mill asked me to start with the company permanently which was very interesting, however there were deterrents to this also, the permanent staff for this company were partners and drank a lot of liquor, occasionally one of them became an alcoholic, then there was also the fact that if you were out in the woods quite a lot it wouldn't be much of a life for a married couple if they had a family, and I figured that someday I would get married. There were other worlds to explore also and I didn't want to settle down yet. The second world war was in progress and I thought of joining the air force as I liked flying, and thought maybe I could help defend our country.

Something I should tell about before getting too far from the bush camps is in the bunkhouses, there was always some rivalry or tricks pulled on one another, sometimes quite serious and sometimes real funny. One fellow nailed his chum's shoes to the floor, another fellow filled someone's moccasins with dried peas then poured water and when the peas soaked up the water, they stretched the moccasins so much they broke the seams. I just don't remember all the things that went on, there was always someone doing acrobats or showing off their strength, and of course the odd braggart, usually some big fellow who needs to be put in his place. I remember one case where this happened while I was a helper on the crawler tractor; this was on a weekend when everyone was I camp. The operator and I were preparing to leave on Monday morning to go out for freight; we always carried an extra 45 gallon barrel of fuel for the tractor, sometimes two depending on distance we had to travel. We would usually place the barrels on the sleigh rack then pump the fuel in, however this day the operator told me to fill them on the ground, so I complained and said we were only making unnecessary work for us, he said to never mind, we have a braggart in camp that can lift them, so I'll get him to put the barrels on the sleigh rack. The tractor operator who was a stocky fellow, very quiet was tired of this braggart and his talking about how powerful he was, went into the camp and called him out, of course he made sure that a lot of men who were in camp that day heard him ask the braggart about giving him some help as he needed someone real strong, so this fellow who was quite a lot larger than the tractor operator, came out to the sleigh where the barrels were, and of coarse there was a gang following to see what was going on. Syd, the tractor operator said "throw those barrels on the sleigh for me," so the fellow walked over grabbing one of the barrels by the edge and almost took the ends of his fingers off, he didn't lift but said "these barrels are full, do you expect me to lift these?" Syd said "aren't you able to lift them?, and" this guy said "no and neither can anyone else," with that Syd went over to the barrels, picked one up and set it on the sleigh then turned to the big fellow and said "we have been listening to you bragging long enough and from now on, keep quiet about your strength," and a few days later this fellow left the camp. I said to Syd later one day "had you ever lifted one of these barrels before?" and he said he could do it that is why he had called the braggart out. You never knew what you were going to see around the camp, as I've said before. One day we were loading pulp racks on the sleigh, preparing to haul pulp, and each teamster with a team of horses that was hitched to a sleigh would pull up in front of the racks and several men would lift the rack on; one fellow who was called Shorty came up with a real snappy team of horses, the team was feeling good and ready for excitement, which they got. Shorty dropped his lines down to help with his rack and off the team went, on their own, they ran behind the office where there was a double holed toilet, so when the team started off really quick, the tongue on the sleigh dropped down between the horses, they ran straight to the toilet and veered off really quick, but the tongue went straight ahead through the middle of the toilet and as the horse swung, half the toilet came off with them, and they went on around the blacksmith shop, and out of the other half of the toilet came the old scaler as white as a ghost and pulling up his pants, he headed for where he slept, and he said afterwards he didn't have to go to the toilet again for several days. He was real fortunate he was sitting where he was instead of the side of the toilet that was peeled off.

One morning the milk train arrived late for some reason or another, I'm hurrying to get the milk pasteurized and delivered, there was a faulty main breaker at the pasteurizer and sometimes the breaker would cut out. This was the old manual type which we never see any more. It was in a box with two bars attached to a handle outside the box. When you pushed the handle up the bars are supposed to go in between four bars at the top of the box, the four bars were always live or where the electricity came into the plant. One of the bottom bars were slightly bent, and sometimes did not go in all the way and would kick out when the power load became heavy. This particular morning I don't know why, and especially when I was running late, the breaker kicked out several times, finally I went over, opened the box, reached in and squeezed the bottom bars together to straighten them, I had rubber boots on, was standing in usually ½ inch of water on the dairy floor. Somehow accidentally the power from the top bars must have touched my hand and as I remember, I was thrown in a circle with my arm held in a curve and everything turned green, and out went the power. I didn't have any electricity to pasteurize with and sure didn't feel very good so went in the office and laid down on a couch for a while, and thanked the Lord for still being alive, and eventually the power came on sometime that afternoon, so I got by and finished my work just in time for my supper.

The place I boarded was with the hydro's diver and his wife, and they said they had a terrible day at the power plant, everything shut down about 10 am, and they hadn't found the problem, and he said he was down under water a good part of the day checking for shorts or anything tat may have caused the outage; he also went on to say that it would be pretty costly to anyone that they found had caused it. They even called in extra electricians that day. When they couldn't find the problem, they turned the power on again, but all those people they were supplying power to had to operate on a shortage until the second generating station was in full operation again. I never did tell anyone about my experience, not in that town anyway, and I know the Lord was looking after me that day because there had been several killed under similar circumstances.

The fall of 1939 I didn't go back to bush work, instead I started selling an item called a master combustor which was used in coal furnaces. With a master combustor the user could burn slack coal or coal dust and with this sitting in the center of the fire chamber, it would let enough air in that the dust would completely burn, was shaped like a cone, came in 4 different sizes for different fire boxes, which sold for I believe $6.95 to $16.95, and these sold quite easily and I did very well at it. I had a chum who was also selling and we would have a little contest going on between the two of us. One day Jack came in and said he had run into a real tough prospective customer who chased him off of his business. Now here was a real challenge, so I asked him who it was and what had happened, and when he told me, I said "I'm going to see what I can do there." A few days later I called on this fellow who was one of the larger car dealers in Fort William, and he needed 3 of the largest units for the business he controlled, so I went to see him at his office and when he found out why I was there, he proceeded to show me the door. I didn't move, but said to him," you don't know what I'm selling, and I can save you quite a lot of money, you can buy slack coal for your heating instead of the more expensive coal which you now buy." I had done a little homework before I called on this fellow, so he stopped and listened to me and I sold him 3 of the largest units. I think I made 20% of the selling price at that time, so made a fair days pay and had topped my chum, which was all fun to us those days. This product was a new item, was selling real good and the company was looking for representatives to go into new area, and if we took on a new territory, we would get an additional 10% if I remember correctly, so after some consideration, my chum and I decided to go to the Kirkland Lake area where he had played hockey a couple of years, and it was an area that had lots of coal furnaces. We went up there early in the winter and sales went along real well for a while, then the people who owned the patents for these combustors decided to go to a manufacture that could make them cheaper. Well that was the downfall of those combustors. When we sold them, we guaranteed that if they melted down, we would replace them at no cost to the customer. Well wouldn't you know it, we sold a bunch of those new ones from the new manufacturer and had to replace everyone, which also promptly melted down. We were in real trouble and spent more time replacing units than selling and eventually ran out of stock. When we complained about it, we were advised we must have got a batch of faulty units, but they continued to melt down and eventually we had to quit selling, we ran out of money as they were shipped c.o.d. to us less our wholesale price. We spent about six weeks in the area waiting for the manufacturer to get the units o stop melting, but they never did, and what had been a real good product, was now a failure because of the patent holder's greed. Eventually I had to borrow money from my grandfather in order to get out of Kirkland Lake. On my way home, I went to London Ont. to visit an aunt and uncle and some cousins for 2 or 3 weeks. I believe my bus fare at that time was $19.00 in the spring of 1939.

 I went back to work for the trout hatchery again. I worked there until August of 1939, when we had completed distribution early, as two of us were working that year.

 The fellow who owned a large car dealership in Fort William that I had sold 3 master combustor units to came looking for me to sell cars for him, as he said anyone who could sell anything to him could sell cars to anyone, I liked selling so I took him up on this. When I went to Fort William to sell cars, I found board and room with some friends of my dad. He was a French Canadian married to a Syrian woman and she cooked everything the way she had been taught in Syria. I arrived there when the peppers were in season, and we ate peppers every way imaginable for the full time I was there and most of the time they were palatable but I had my fill of peppers for a long time. As it turned out, this didn't last very long as the second world war broke out and everyone stopped buying, and to make matters worse the owner of this dealership had some of his family going to school in Italy, and Italy or Mussolini had joined with Hitler, therefore no one would buy from this dealership while Canadians thought he was supporting Italy. I finally told him after three months that I was going to move on; he said "I suppose you will go and join the armed forces, get yourself injured or killed, and what for? This is the second world war, my brother and I came to Canada before the last world war started, my brother went and joined up to support his new homeland and I stayed here in Canada. My brother has a bicycle business and I own several auto dealerships plus other real estate, so you better think about this; the country didn't give my brother any assistance for his service to his new homeland."

I went back to the Provincial Paper Company in the bush camps, spent that winter as camp clerk and scaler.

I have to return to the period of growing up on the farm; mother and dad were both real good cooks and dad thought that each of the family should learn, so needless to say we all did, as we became old enough to learn, then as each of us progressed, we would have to relieve mother for a day, she wouldn't rest, but it was a change. Mother would go out and work in the garden or find something to do. Whenever you were appointed cook for the day that meant a full course meal not a snack or sandwich etc. I was good at the main course on meals and as mother had preserves, that is what was used, however my brother Wray liked making deserts, and we never knew what he would come up with; there was always one thing, whoever was cook had to have some form of desert for lunch and dinner, or supper as we said on the farm. Learning to cook never did any of us any harm; in fact it became very helpful occasionally through my life.

As I was growing up and thinking about what I wanted to do when

I went out to work, my priority was owning a tourist resort, I liked the outdoors, could sell myself and especially as I worked for the trout hatchery, calling on a lot of resorts the owners had started out as you might say on a shoe string, and many of them were prosperous and had some beautiful resorts in northwestern Ontario They had built their businesses in fishing and hunting, and there was the occasional summer resort that catered to families etc; but not like today's top resorts.

When I was 17 I homesteaded 160 acres on Cavern Lake where I planned on having a resort some day, it was a very scenic area before all the timber was cut, and there are canyons and lots of lakes in the area.

I recall one owner who had a small resort, he was a bush pilot and took fishermen into remote areas and he was also the man that invented the skis for float planes so they could be used in the winter I think he made more on the skis than at the resort. I did finally own a resort in later years which will be in more detail later on in this biography.

When we moved to my grandfather's farm, we had two neighbors, one across the country road, there was a teenager about five years older than I, he and I went on some of these toboggan slides I talked about earlier and of course we walked to school together until he had finished public school, then the second neighbor was a sawmill owner with their quarters for office, cookhouse, and room for the employees all in one building. A year or two after we moved to that area a family moved to the mill where a house had been constructed, he was mill foreman, there were 3 girls and later one boy. We of course got to know them as we all went to the same school, and also had a lot of fun going to and returning from school in the winter, it would be snowball fights, the three girls were about a year younger than myself and each of my brothers.

As the years sped by, I began to realize that the oldest girl was quite interesting to me, and from the time I was about 16, she was my girl friend. We went to all the parties, dances, and activities in the community together, and we did look quite a bit alike and once or twice we heard comments from older people who didn't know us say "isn't it wonderful how that sister and brother get along," and we did get along really well. We had lots of discussions about everything, were not jealous of one another and even had it arranged mutually that if either of us met someone at a party that was interesting we could take that person home or go with them whichever the case was, however if I met someone, I would make sure that Fran had a ride, however I don't think that ever happened as she was very popular and a good looking girl. As time went by we went together permanently and planned on getting married. When this finally came down to seriously getting married there was one problem that arose about religion. Fran was Catholic and I was Protestant but she couldn't leave her religion or she would have been disowned and this may have happened to me although my family was more open to the idea. I said I would try going to her church, and by this time the mill had closed due to a fire, and her family had moved to Port Arthur, about 45 miles from my home although by now I was away quite lot wherever I worked. We decided to get married in about a year when I would have been 21. I gave Fran an engagement ring and at that time I was working in Port Arthur and went to her church, also the church bingos, which brings up an interesting anecdote. We were at a church bingo one evening and I won, and so did another lady who was one of the leading ladies of the church and area, and the winner could have a turkey or a basket or groceries, because there were two winners the master of ceremonies asked if it would be alright if we cut a deck of cards to see who would take the prize, it didn't matter to me and as it was the catholic Church and as she was a Catholic it didn't matter to her. We proceeded to cut the cards, I offered her the first choice as I thought ladies always have the first choice, however she said "oh no, it is unlucky to cut first," with that I cut the cards and as I did I also said "here is the ace of spades" which it was to my surprise, I had said it as a joke, so the master of ceremonies looked at me and said "did you ever do that before," and I said "no I'm not a professional card shark," that lady took off down the hall in a huff, and I took the basket of groceries, which I gave to Fran for her family, but it was all done in fun, and we had quite a laugh about it.

    As it turned out, the work I was doing at the time, selling master cylinders took me away from Port Arthur to northern Ontario, and I continued to go to the Catholic Church while in that area, but it was so much different from the Protestant Churches I had attended that I started to have different thoughts and started analyzing the situation. Fran's father had been a Protestant before marrying her mother and he was not a very happy man, never spoke of his parents or of where he came from, and I started to think I could be the same some day, and

I didn't care for the type of religion, although I loved Fran, I began to rationalize our situation, and finally came to the conclusion to put this all down in a letter to Fran, which I did, and also said that if she couldn't change that we should maybe reconsider marrying, but she would always be a good friend, as it turned out that is the way it is. We had made plans that when we were married Fran would go to work and I would go back to school and get a professional education as we each thought we could improve ourselves financially, and growing up in the hungry thirties, that was uppermost in our minds, although each of us had faired rather well. The good end to this story is that Fran did marry a real nice young man who was a Catholic 3 or 4 years later and she worked while he went to school specializing in electronics and plastics, he later was one of the engineers in the development of the atomic bomb in the U.S., and still live there, but are good friends of ours. Fran's husband had been quite friendly with my wife when they were going to Tech.

After I returned to Port Arthur in 1940, I would go out with Fran occasionally but were careful not to get too serious, however one time Fran and I were out together, she told me she had met a nice Protestant girl I should meet, liked the outdoors, fishing, hunting, canoeing, and would probably make me a real good girl friend and maybe a wife; Fran was still out for my welfare, and eventually she did introduce me to Merelie who later became my wife. Fran had prearranged to have Merelie and her brother Verne meet to go to a show or whatever, but it turned out after meeting, we did go to a show, only Fran and Verne went to one theatre and we went to another as Merelie had seen that show earlier that day. While walking home that evening, we were going through a back lane to Merelie's house and a cat came scooting across in front of us and Merelie almost jumped on my shoulders, the cat startled her, she was embarrassed, but she said she was frightened of cats which was quite apparent from my point of view.

We continued to go out together, became real good friends and soon became quite serious; I only went out with Fran once after she introduced me to Merelie, she told me she was becoming quite interested in the fellow she finally married, and we decided to each go our own ways but always be friends.

Merelie and I met on Friday July 18th, were engaged on Monday October 14th, and married November 30th 1940. She was 18 and I was

23, and was married in Merelie's family home. Her aunt and family were there and my family and grandfather, it was a small wedding. We had a real short honeymoon, went to one of Merelie's aunt's camp on Lake Superior for two days as that was all the time I could get off. There was about two feet of snow, the bay was frozen over so we went skiing, and Merelie's brother Verne and a girlfriend came down to the camp on Sunday and we spent the day together. Merelie made a chocolate pudding which we will never forget about, it was large enough for 20 people and we have often laughed about it. We lived with Merelie's dad and brothers Verne and Ray. Merelie's mother had passed away in August with a massive heart attack.

Verne, Ralph, Merelie and Marg

While going out with Merelie, we spent a number of weekends at their summer home on Island Lake in the Lakehead area, a beautiful spot which is still in her family. We would walk, swim, and loaf around and canoe; Merelie was a real good canoeist, in fact she and her brother Verne had won a number of trophies for canoeing. She could pick up a 15 foot canoe and put it over her head for portaging as easy as I could. One evening on Merelie's 18th birthday, August 10th, we canoed up the lake a ways, pulled in on a point and were watching the stars as we were talking, we saw a star fall and counted 18 fall while we were there, and have later learned that a lot of stars fall in August and September.

We would go to a party or dance, stay to the very end and then go fishing somewhere, quite often around the Dorion area where my parents, brothers, sisters, and grandpa still lived. We have seen more wild animals when going fishing early in the morning, which was usually just at daybreak, we both really enjoyed this type of life and always have over the years.

When we were married, I was employed by the Port Arthur Ship Building Company as a carpenter, or at that work, I was commonly known as a woodworker. I had started there in the spring of 1940.

After leaving the bush camps that same spring, I had tried to join the air force, but had been turned down due the back injury I had while working in the bush about 3 years earlier, the injury didn't bother me and I didn't think I had a problem, however the doctors rejected me as they could see it in my back when I walked. I then went to the shipyard and started there to help out in the war effort. That same spring Merelie's brother Verne started to work there as a draftsman and later joined the air force where he became quite outstanding and won some distinguished medals.

When I went into the shipyard where I learned to glaze windows, I was quite adaptable to the woodworking equipment, and as is the expression as adaptable as a duck to water. I must tell you about when I first started there. You had to go to the superintendent and he interviewed you and if you were hired, he told how much your wages were to start, in this case he said they would hire me and the rate was 35c per hour. I soon figured that out in my head and told him "no, I wouldn't work for that wage, it had to be a minimum of 45c to start and I would expect an increase when they saw that I could produce," if not 45c, I could go back to the Provincial Paper Company where the bottom wage was $4.00 per day at that time, the superintendent was surprised at my reaction and said so, however he agreed to the 45c rate. You were paid every two weeks, and when my pay cheque came through, which was always on a Friday about an hour before quitting time, I figured out my time and they paid me at 35c, I stopped work and went up to the superintendent's office, gave him the cheque and said if the company couldn't afford the 45c rate we had agreed on, they had better keep it and I would go to work where I knew the pay was better; he apologized for the error and said to wait until Monday and he would have it rectified as I was a good worker and they wanted me to stay. Monday he brought the cheque to me with the proper rate as we had agreed on, however I don't think they had ever had an experience of that nature before, I then decided that they tried me, and from then on I never stopped going after more money while in their employ, in fact when the carpenter union came into effect about 9 months later I was drawing 87 ½ cents per hour, 2 ½ cents more than the labor union rate that was set for work performed while there.

That was an experience, the I.W.A., the International Woodworkers of America, I was one of the employees that was endeavoring to get the union in the shop; I was of the opinion that a lot of their employees should have received more pay. No one knew what they were paying me as I was told by the management that they would give me the raises I was getting on one condition, that I didn't tell anyone else. There were also lazy employees who just filled in a space; I thought the union weeded these lazy employees, don't kid yourself, the lazy employees were protected by the union, I couldn't believe it. The most startling part of bringing in a union was the ones who weaseled their way into the control of the union, as far as I was concerned, they were lazy workers and were communists. Shortly after the union was formed by the executive body that took control of our union, although I had been instrumental in bringing the union into the shipyard, I would have nothing more to say, and when I objected, they were going to throw me out. I soon learned there was a small group who controlled the I.W.A. and I regretted doing what I had done. I did continue to belong to the union and that became a closed shop tactic, if you didn't join you didn't work, and under the union shield, the lazy ones were protected.

While I worked for the Port Arthur Shipyard, my first project was glazing windows. In the days of wooden frame windows, the glass was inserted in the frame and then it is sealed in around the glass with putty which seals out the rain and holds the glass in place. There is real art do doing this type of work, known as glazing, and I had a real good teacher and he held one of the top records for glazing so many windows per hour, but I don't remember the number at this time, anyway I became very good at it, did this for about 3 months, then went on the woodworking machines, which was very interesting, they took on a subcontract of building Hawker Hurricane airplanes for the Canadian Car Company, and I was transferred to the manufacture of planes. I was bench foreman on the emergency door exit panel. These planes were made of Sitka Spruce veneer. The only protection the pilot had was the ¾ inch Plexiglas in the windshield, and these planes were considered the fastest thing in the air, they could travel at 125 miles per hour from point A to point B, or the turn of your body, 0 to 125 miles was really something in 1940; these planes were going to be fighter planes for the war. I only worked for the shipyards for about 10 months and then went into business for myself.

Some years later when I hadn't been working on any project where the union had any say, probably for me, I recall I had been ill and my union dues had not been paid. When I was able to go back to work, I went in this particular construction site one morning looking for work, there was a line up of probably 50 men, the superintendent came down the line and picked workers which included me; I had worked for this superintendent before and he knew I was a good carpenter. After I had been working about 2 hours, the shop steward came to me and said "your dues are not paid up, so unless you pay them up, the union is going to strike unless I was off the site by noon." I immediately went to the superintendent and told him about the steward's action, the superintendent said "we will pay your dues," and I said "no, if this is the way they treat their union members and especially when I had been ill, I would not pay any more dues, would go back to work for myself." The superintendent said he didn't want to see me leave and offered to send me to Edmonton where they had another project under construction and there was no union. This meant I would have to move my family at that time, better to stay at the Lakehead where I knew I could work for myself, I thanked him for the offer and explained my position and never did pay any more union dues and have never had much respect for unions because of the type of people who are usually managing them, they have their place but there should be better control and consideration for their fellow workers. The shop steward went to the superintendent that morning and said "there are lots of men out there looking for work why not hire one of them instead of me" and he replied "most of those bums I wouldn't have on the job and that is why they are not working for me. I had them before and I don't want them, the union should weed them out instead of protecting them." the superintendent was representing Dominion Construction Company, one of the largest construction companies in Canada. I had worked under the supervision of this superintendent some years earlier on a grain elevator at Port Arthur and on the construction of a paper mill at

Marathon. He was a tough but yet good fellow to work for if you did your work. When I first started working under him, I would hear him shouting at men and I worked several days just waiting for him to shout at me and I would have told him where to go quickly, when he didn't shout at me, one day we really had been going steady and hard, as this was on a site where they were pouring cement and the carpenters had to keep ahead of the cement as there were times that the cement was poured non stop, depending on the building. When the cement slowed down, I purposely took off my apron, put my tools down so he could see them and went down one level for a rest, and in a few minutes he came down and looked at me for a few minutes, then said "taking a rest," I said "yes", and he said "you need it after that last pour" and we had a little talk about the project, then I went back to work and he never did shout at me, so I started watching who he yelled at and it usually was someone who was sloughing off on their work or didn't know what they were doing, and after that, I considered him a good construction boss. I did find out later the company considered him one of the best with very little injuries on the job. Now speaking of injuries, while working on the Marathon Paper Mill project, it was in the winter and very cold. One morning while framing, the head of the nail slit and hit me in the eye; I had to go to emergency right away, the first aid attendant put some freezing in my eye and sent me back to Port Arthur to a doctor. Marathon is about 200 miles east of Port Arthur, and had to go by rail, as it turned out I was on compensation long enough that the project was almost completed, so didn't go back.

This type of work was usually short projects and most of the time I would go on one if I didn't have something for myself to do. It would be good for a change for a short while but I was always glad to get back home.

Back to when we were married, Merelie's family usually had a maid while Merelie was growing up and had learned very little about cooking. It didn't take her long to pick it up, although she never did like cooking and would much rather be out working at the projects we endeavored and almost all the time together in some form or another. Shortly after we married, Merelie decided to make a spice cake for dinner that evening, this is my favorite cake, so after the main course of dinner, she brought out the cake, it looked beautiful with a nice white icing. Merelie's dad took a good sized piece, and I noticed his face seemed to brighten up and I assumed he really liked the cake, so I also took a good sized piece, and once I lit into the cake, I certainly knew it was spiced and why dad's face brightened up. Merelie had warned everyone that if anyone criticized her cooking, they would be the next cook, so needless to say dad had the second piece, but I couldn't, and tried to get Merelie to have a piece, but she said that wasn't her favorite cake and she had made it for me and I only had one piece, so finally I had her try a small piece, and when she tasted it, she said "wow, how did dad take the second piece?" As it turned out, Merelie was following the recipe, it called for this and that type of spice including allspice, not knowing that was a certain type of spice, she thought it meant all of the spices, so she put some of every type of spice in the cake including poultry spice and cayenne pepper and this really made that cake hot, needless to say the cake went out, and Merelie had a good lesson on cooking, as we laughed about it. She did say when the cake was cooking, there were a number of different odors coming from the oven.

Merelie and I had several things in common although we had not known either of our families previous to our meting, for instance, we bought our marriage license from my grandfather, my mother and dad had bought their marriage license from Merelie's grandfather 24 years earlier, our family names were similar, my name is Ralph, and Merelie's brother was Ralph, my second brother is Wray, Merelie also had a brother Ray, then there is my third brother Lorne and Merelie's brother LeVerne, and I have a sister Marion. There are also other family names such as Merelie's mother's name was Electra and I had an aunt Electra, and we each had an aunt Eliza. Merelie's mother was given the name Electra by her father who had been instrumental in having the electric street cars established in Port Arthur, and she was born on the first day the street cars started running on the 29th of March 1892.

When we were married, dad and I still owed quite a lot of money for debts from the dairy that was burnt, so we didn't have any extra money to go out very often, so we would play crib and listen to the radio, and on pay days we really splurged by buying two chocolate bars and a quart of milk and another good game of crib or two. Merelie doesn't like playing crib any more to this day.

We were not tied down too long, eventually bought our first car, a 1927 Chevrolet sport coupe for a few dollars, it needed some body repair, I remember making some wooden parts for the body, but we made it look quite presentable and were real proud of it, at least now we could go fishing and out where we liked to be but this was not my first car. I had owned a 1934 Pontiac sedan in 1937 while employed at the Red Rock Pulp Mill Construction, bought it with a small down payment from a G.M. dealer in Port Arthur named Rothschild.

Back to Merelie, we used that little car for a while, and then Merelie's brother had what we thought was a better vehicle for sale. It was a Willys Victoria coupe about a 1930 model built like a tank, large wheels, ran quite good most of the time, and it did stop one time while coming back from a fishing trip, however no fault of the car, this time it was the driver who tried to cross a flooded highway and the wiring got wet and we stopped right in the middle and had to be towed out. Eventually we got rid of this monstrosity and bought a little red coupe about a 1930 model and Merelie really liked that one and really had a lot of fun in it.

Merelie really drove fast although she was a very good driver, very sharp but sometimes took chances, one time almost cost us our lives. She stopped at a railway crossing for a train to go by and as soon as the caboose was clear, she stepped on the gas peddle and sped right in front of an engine of a train on the second track, we were so close to that engine it rocked the car, however it was a good lesson which we both lived to talk about and be much more cautious at railway crossings.

Merelie was always full of mischief from the time she was a child, so some of the things I relate, you will gather this. Her second name is Mae and we used to say "Merelie Mae, and then again she may not," because she wasn't sure what she would do until it happened. One time when she was about 6 years, her parents were moving some household items out to their summer camp, they had to cross a lake with an old flat bottom boat, and sometimes the wind would come up suddenly and could cause the boat to take on water especially if it was overloaded. This particular day, it happened and Merelie's brother Ralph was running the outboard motor, the waves were getting too high and he said "we may have to throw some of this stuff overboard to lighten up," and before anyone had a chance to say what would go, Merelie threw the canary cage over and then some large pictures with frames. I said to her one time she was telling about this, "what did your mother say?" and she replied "I don't know but I never did like that bird." Another time she put pepper in the bottom of every chocolate, in a box that her mother had for treating guests, and of course sat around to watch the faces of the chocolate eaters, she said "I really got in trouble that time."

In February 1941 I was still employed by the shipyards when I heard of a position that was open for the newly formed Government department, war time housing where they had an office in Fort William, so I took the day off and went to see about it. The position was purchasing agent, which meant calling on all the mills in the area that were sawing lumber, and purchase whatever lumber was needed for the construction of houses for employees that were working on Government war projects in the cities. When I went in the office and asked about the position, the gentleman doing the hiring was about 60 years, he looked at me who looked like about 19 or 20 maybe, and said "I think you would be too young for this position and besides you might be called up for army service at any time." I assured him I had been turned down for the air force and also the army. As for the army, I had been called up the previous fall, but after a route march from the army barracks in Fort William out the Scott Highway for the day on the paved highway, I couldn't get out of bed the next morning because of a pain and partial paralyses in my back, now I really knew I had a back problem. The old sergeant stormed into the barracks where I was in bed and yelled "get out of bed and out on the parade grounds," and I assured him that if he could get me out there I would be glad to, in fact I got him to help me to sit up then gave me a note to go to the doctor whenever I was able to walk there. Unknown to me, some in the camp had measles and had them before, but every time someone had measles and I was near them, I would get them also, however after being in bed for a day or so, I finally did get in to see the doctor, after I was examined, he said he was going to give me a weekend pass and I could go home, so Merelie picked me up with our car and home I went, and what do you know, the next morning I was covered with measles and had to stay home until I was over them. When I did get back to the barracks, they promptly sent me home as they said I would be discharged due to a back injury which I had told you about that happened in the bush several years previous. That back injury which would cause me some pain and problems later on had stopped me from being another war hero or fatality, whichever our Lord had willed; in this case he saved me for other adventures.

Back to the position with war time housing, I told him about my background and my age which was just two months short of 24, he looked me over and said "I haven't anyone else at this time so I'm inclined to give you a trial for two months," and proceeded to tell me the outline of what I was supposed to do and also the wages. I would start at $265.00 per month with expenses for traveling, and in those days that was a fairly good wage. I thought about all the thousands of feet of lumber they were going to need and all the mills that were scattered around the Lakehead area, and a thought came to me, why not work for myself, buy the lumber needed and then sell it to the war time housing. So I said to this gentleman, "do you mind if I make a suggestion and ask a few questions? and he said "of course," so I asked how much the department would pay for a thousand feet of lumber delivered to their building sites in either Port Arthur or Fort William and he said they were prepared to pay $37.50 per thousand. I knew that I could buy a lot of lumber for less than $20.00 plus the planing and hauling, so I knew immediately that I could make some money here, so I replied, "how would it be you don't pay me any wages, but give me a letter stating my position of purchasing lumber also that I would have all the gasoline I would need plus a permit to purchase tires, (there was rationing on gasoline and tires at that time) for whatever vehicle I used?" He looked at me and stepped back a pace or two and said "you must know something I don't or are smarter and more adventuresome than most but I'm willing to give you a try." I left the office with the letter and permits for gasoline and tires and on a two month trial, boy was I happy; this was a chance in a lifetime as I saw it.

I went to one of the larger companies that had a planing mill, Northern Wood Preservers, and arranged to have the lumber I purchased dressed and a place I could stockpile the lumber as it was delivered from the mills, also for trucking the lumber, sometimes from the sawmills and also from the planing mill to the location where wartime housing wanted the material. prior to this, I had gone to the bank and arranged for financing this operation as I didn't have any money and in fact I had to purchase a better vehicle shortly after starting, so this time we bought a 1934 Plymouth sports coupe, a real flashy looking vehicle for those years which was our fourth vehicle.

I really went to work, calling on every mill I could find, and for a while I arranged for 90 days on payment of lumber, 90% of them were small farmers who would cut logs on their farm during the winter and operate the mill in the summer along with whatever farming they did. There were a few who had year round operations and some had quite a lot of lumber stockpiled just waiting for a fellow like me to come along. These were the ones that supplied a lot of the lumber to start with and by the end of May 1941, I had the original planing mill overstocked, and therefore I would have to find another planing mill as they could not plane ours and do their own work.

I had been approved by wartime housing and therefore could start thinking about something more permanent. Merelie and I were working together, in so far as she would look after some of the accounts and did tallying of lumber that was shipped from some of the mills, so I would drop her off at some loading or mill site and go on calling on mill owners in the area for the day. After some consideration I decided to try to purchase a planing mill, and this was made more realistic when one mill I tried to get to do planing for me wanted quite a lot more than I figured I should pay.

There was a railway siding on 30 city lots which would make a good lumber yard site and a place to ship from. I arranged with the railway and the city to use these lots in the city of Port Arthur with an option to purchase the lots, and then went out looking for a planing mill. I also had to find someone to operate a mill and these types of people were very scarce, and I knew of only two and both were employed, anyway, I offered one fellow a deal he couldn't turn down and he accepted; he had been employed at the shipyards in the same department that I had formerly worked in but had a lot of sawmill experience which was the type of fellow I needed. First I offered him a good wage, better than he was getting but this didn't appeal to him so then I offered him a share in the business, such as half interest in the planing mill which I would purchase and he would operate, but I had to have someone dependable as I was out traveling and purchasing timber full time. He wanted an agreement drawn up to that effect which was alright with me, also a rate of $2.75 per thousand and he would hire the men he needed. I went to a lawyer who was acquainted with my grandfather whom he said he thought was quite honest, explained the deal I wanted drawn up, that it would be 50% of the mill operation and not any part of the purchasing and sales of lumber, so he said he would draw it up as I wanted, so I went back out to this fellow to locate a mill, which we finally did, had it set up on the site with a 50 h.p. electric motor for power and it was a nice operation. Shortly thereafter the lawyer called to say he had the agreement ready to sign. I went in one morning in a hurry, asked if it was drawn up as I had wanted and he said it was, so I signed it and made my first mistake in this business. I should have read the contract over; I didn't and lived to regret it.

Up to this time we had originally lived with Merelie's dad and brothers until about the time we went into the lumber business, then we rented a small apartment for a while and with both of us working this wasn't the best. We then moved in with Merelie's aunt where we paid board and rent and our meals were prepared for us, so this was okay for a while until we took George Johnson into the planing mill and when George and his wife Ina suggested we live with them and could discuss the operation of our business in the evenings and plan together. They were a real amicable couple and lots of room and Ina liked cooking and they had a little girl about 3 years old which Ina had to be home with, this worked out real well for quite a while.

We were expanding and also buying stands of timber and contracting sawmill owners to come in and saw the timber for us. We had also expanded our planing mill operation to loading and shipping railway cars of lumber east, west, and south. The wartime agency had a man in Toronto who we could call or vice versa and request car loads of lumber to be shipped wherever, also about this time we were approached by an American company who wanted us to supply the Americans with material suitable for crating bombs in which turned out to be the cheapest material at the time, which was black poplar or the common name cottonwood, a real tough stringy material and when running through the sawmill, smelled like a sour pigpen, however we were glad to have a market for this product. Some time later during the war, we also sold white poplar to the Americans for the soles of shoes, which they were going to use due to the shortage of leather and rubber etc.

Occasionally we would have an odd request for lumber, like birch planking for furniture. One time I had two car loads of planking that had been left lying on the ground too long after it had been fallen and cut into logs; the birch had mildewed and instead of being white it was light shades of mauve, and I could take a 3 inch plank, drop it across my knee and it would break, so I told the mill owner I would see what I could do with it, but wouldn't buy it until I had the right buyer, and he gave me the phone number of a furniture company in Winnipeg, so I called and told them what the material was like, also that I wouldn't ship it until they sent a representative to approve it, which they did, he was a small Jewish fellow and was real tickled to get this mildewed birch. I showed him how easily it broke and he said that didn't matter, they used it for filler between two pieces of veneer and no one would ever know any difference, so he paid the top price and contacted me later to see if I could get him any more.

Shortly after getting our planing mill in operation we had a chance to purchase a new 3 or 4 ton Dodge truck, these new vehicles were really scarce, we needed one so we bought it, although we did have hired trucks moving material, this would make some work much handier.

One project that came up was supplying squared timber for the construction of grain elevator expansion; in fact we supplied enough timber to double the storage of the grain elevator at the Lakehead. Due to the war, there wasn't enough shipping facilities to move the grain fast enough, therefore had to enlarge the present storage space.

The lumber business went along very well, exceptionally good in fact; I believe Merelie's and my personal income was $13,000.00 the first year, much improved from $56.00 per week the year before. We really enjoyed ourselves with parties and made up for the time we had missed in our first year of marriage. Now we were in to our second year of operation and decided to buy a house for ourselves as Merelie expected to have our first child in the fall and would have to be home, and as it turned out, we rented a house on Pine Street and my parents moved in with us, which was quite helpful. Our son Keith was born on October 21st 1942, a joy to the world, or at least to the whole family, and after recuperating, mother looked after Keith and Merelie went back to work in our business as she was needed and enjoyed being out working with lumber. Mother and dad had lived on the farm up until moving in with us. That was near the time that dad had gone back to cooking and mother was alone at Dorion, so this was much better for my mother and us.

As I have said before during the winter the lumber supply would slow down during the winter months when the small mills would stop sawing lumber and cut their logs for the following summer; there would be less planing and work for our tuck, so we took on hauling pulpwood from one of the pulp companies. I recall one trip I went in with George Johnson, as he said he thought the ice was too soft to haul on, which I readily agreed with. When we came out with that load, we were out on a lake a quarter mile from shore, driving through two feet of water, the ice was spongy and as the load passed over the ice it would settle and the water would be all around the truck, a rather startling feeling, so needless to say that was our last trip in there. We then went out with the truck on another haul where there were no lakes to cross as it was getting close to spring. On our first trip on this haul, I went along to see the type of roads and almost was run over by our truck; there was an ice section on a side hill, so on our way out with our first load, I suggested to George who was driving, that he stop and I would go ahead and cut a groove in the iced area for the upper wheels to run in and this would prevent the truck from sliding off the road, this I did, and when I had the groove chopped out, I motioned for him to come ahead. I stepped off the road on the upper side of the truck, and as the truck passed me, my feet slid out from under me and under the truck I went, I rolled and crawled fast enough to keep away from the rear wheels and finally rolled out on the opposite side of the truck. When George stopped a ways ahead for me to get in, he said "where did you go? When I told him what had happened he couldn't believe it, so fortunately the truck was moving slow or it would have been curtains for me as George never saw me slip.

When spring rolled around, we were as busy as ever in the lumber business.

Along about mid summer one day when I was at the planing mill George had his brother-in-law there and George suggested that we take him in as a partner. I asked George why, and he said he wanted him in; as it turned out I knew his brother-in-law from a previous experience, as he had the franchise for the master combustor when I had been selling them in 1938. I didn't have any special feeling for this fellow as I was sure he had been partially instrumental in changing the manufacture of the combustors, and certainly not helpful with me when I had the problem in Kirkland Lake. Anyway I suppose we could have given him a contract on his own with a sawmill, but I just didn't trust him, so therefore I said no and went on my way, back out purchasing lumber. The next time I came in to the planing mill, George informed me he was finished with our partnership and that he had sold the planing mill, so I asked him how he could sell it, and he said "I own 50%, so you can't stop me." I said I could pay him out, but he said "It is already sold and I have been paid for the full mill which I shall keep as income that is coming to me," which no doubt he had earned. I may have been able to stop this deal but I would have been in another predicament about finding a mill operator, and the business was getting much different from when we started. A lot of the lumber was being sold rough or not planed and I could dispense with the yard; however in time to come I found I should have kept that mill and yard, probably would still be in the lumber business.

I asked George what he planned on doing; he said he and his brother-in-law were going out on the Alaska Highway to work, so I wished him well. This was not the end as far as George was concerned, and a few days later we had a call from Inez, George's wife asking where he was, as she thought he and I had been away somewhere looking at timber or something. When we told her what we knew, she then revealed that he had mortgaged the house for all he could get, sold the furniture and there was someone there to pick it up and she knew nothing about this. We told her what he had said, that he and Ted, her brother were going out on the Alaska Highway to work. We couldn't believe that George would do such a thing to his wife and daughter, as Inez was a real nice woman, a real congenial person with a real nice daughter, they were married and we thought they were getting along real well. We wondered if George was making much more money than he ever had before and his brother-in-law saw this, so talked George into gathering all his money together and going into some type of business together, probably trucking on the Alaska Highway. George drew from our business account $5000.00 unknown to me when he left, and this is when I found out that I should have read our agreement over and not trusted our lawyer, as when I talked to him about it, he said he thought that George was supposed to have 50% of everything, and I should have checked this agreement earlier as I did find out that George and the lawyer belonged to the same lodge in Port Arthur.

When George drew the $5000.00, I had been writing cheques to some of our suppliers and when there was one or two cheques n.s.f., I had some explaining to do, almost lost some suppliers as the word travels fast and the word was, I was going broke, it didn't help business and besides by now there were other lumber buyers out looking for material now. Number one lesson, never let anyone have a 50% interest in a business venture again. Along with this problem, the Bank of Commerce whom we were dealing with changed managers at this time and cut off our credit until we evaluated, he said. When I got in to see the new manager, I was a little peeved to say the least, so I asked him if he had checked to see how much money was due to come in, as up to now, I had all payments for lumber sold, payable to our account at the bank there were two reasons for this. When we made an invoice for lumber sold, a copy went to the bank, they always knew how much they could advance if we were overdrawn, secondly, at the time, the income tax department didn't check the bank accounts, then there was another reason, we were getting along without an accountant which I soon realized was not the right way to operate a business. Due to the way the bank had caused a problem also, we changed banks a couple of times before we found someone we could depend on.

I have to revert back to something that happened a couple of months after Keith was born. When the lumber business slowed down in the winter of 1942, the forestry department had enquired if I would be able to do some check scaling for them, I asked for how long, and as it was for a short period I said "okay" and went out to a camp in the Marathon area again, about 200 miles east of Port Arthur, it was into a bush camp about 35 miles from the train depot, I went in most of the way by truck, the last 12 or 14 miles I had to walk. When I got into camp, I went into the cookery for a cup of coffee and doughnut or whatever, and while in the kitchen, the cook opened a box of raisins, and I like raisins so I had a handful or two, the cook said "you better be careful with those raisins they are frozen," well the result was I got sick from eating those frozen raisins and thought I would die, then for a while was afraid I wouldn't, anyway I survived and after a day or so I started check scaling, but it took longer than I had planned and now it was getting close to Christmas, and anyone that was going out had to be at the next camp on a certain morning by nine o'clock. I had to walk to the next camp, and just before I arrived I heard the truck leave, now I had to walk another 25 miles to the railway and be there before 4 p.m. to catch the train which ran daily; there was no place to stay if you missed it, only a freight shed which was mighty cold by that time of the year. Well anyway I walked out, had some company following part of the way, there were a lot of timber wolves in that area, sometimes they would be quite close and sometimes back far enough to be just visible, however I didn't stop to enjoy their company. One of the tractor operators claimed he had been attacked by some wolves earlier but I had been near wolves quite a few times and had not had any bad experiences, except that they liked to follow, and would sometimes give me a creepy feeling up my spine. Anyway I arrived at the railway just shortly before the train, got on board and what do you know, the passenger cars were all full, there was standing room only, I guess everyone was going home for Christmas and I certainly would like to sit down after that walk. Well I stood up most of the way, arrived home about 2 a.m., Keith woke up and started crying, which he very seldom if ever did before or after, and I walked the floor with him most of the night. I never forgot that trip and Keith was too young to remember it, so every once in a while I remind him now just for fun.

We had a real good Christmas that year, with all the family home, everyone had a bottle of cheer to my mother's dismay, we allowed liquor in our home, had it been at mother and dad's it would be different as mother did not approve of anyone drinking strong beverages.

Before straying too far from George Johnson I should mention a couple of interesting incidents. About six months after George left, Inez received word that he had been killed in an accident on the Alaska Highway, they had found his wallet with her address and said that his body would be sent back in a sealed casket, which it was, apparently, as no one opened the casket to see if it was him, however in 1956 I was driving down one of the streets in Vancouver B.C. (Trafalgar) and I felt as though someone was looking down my neck, I turned my head and in the next lane in a large black car was George Johnson driving beside me and he was looking at me, I couldn't believe my eyes, he realized that I recognized him and he took off and I couldn't get his license number and eventually lost him in the traffic; so I do know one thing, it wasn't him in the casket, but I've never seen him since either.

In 1943 I was still buying and selling lumber, however the wartime housing wasn't requiring very much and most was loaded in rail cars and shipped out of the area including the U.S.A. There were others crowding into the business and offering much more than I would pay, in fact one fellow started up in business and borrowed himself into debt that the bank couldn't cut him off credit, he started with nothing and had nothing to foreclose on, he told me he owed $150,000.00 when the bank realized their position and had to carry him until he could work his way out, said he had done it intentionally. He was paying so much for the lumber that it was at times as low as 50c per thousand board feet between purchase and selling price. By this time, had I kept the planing mill, I would have been well established and could have gone into retail also.

We had moved into another house on a rental purchase arrangement, a nice little bungalow on Martha Street in Port Arthur and carried on purchasing and shipping lumber. Whenever I needed Merelie to tally lumber when loading box cars, mother would take Keith which she really enjoyed, in fact he spent quite a lot of time with mother, dad was out cooking and it was company for her when my sister was out working too.

Our truck which we didn't need for the lumber, only occasionally, we had a driver on it working for a contractor which was building a breakwater for ocean shipping to come into the port at Port Arthur, and as it turned out, it wasn't the most profitable situation as we lost three dump boxes while the truck was there but was bringing in some income. While our truck was there, we bought a large supply of lumber from the Leahy Timber Company and had to have it hauled to the railway for loading, so hired a tandem truck and trailer with driver. He wasn't the best driver and was not getting the material out fast enough, so one day Merelie and I drove out to the site to see what was taking so long. Well to begin with, he couldn't back up a truck with a trailer and spent an hour or more trying, so eventually Merelie said "I can back that truck into there," so I said to go ahead and try because I wasn't sure I could, so she said to the driver "let me try that," well the result was she backed the truck and trailer in on the first try; the driver was so peeved he quit right there, and to make matters worse, the loading crew had to leave as it was their supper time, and in the bush camp you have to be on time for meals or you don't eat. Merelie and I loaded that truck and trailer, and then proceed to take it out to the loading area. We had ridden in with the driver and truck so had to take the truck out in order to get home that night. On the way out, I was driving when I hit a chuck hole and broke a rear spring on the trailer, the load settled down on the tire and there we were. As it turned out there was no jack with the truck so now what do we do, it is dark and not very likely that anyone will come long that road until morning. We had an axe or two with us so decided to make some birch wedges and pound them in between the spring and bunk to lift the load off the tire, and this we did. After we had it raised up high enough to travel which had to be slow so the wedges would stay in, we proceeded on, however when we got back in the truck, I noticed Merelie had something tied around one finger, so I asked her what happened and she said "oh I nicked my finger when we were cutting the wedges." I wanted to see how bad it was but no way would she show me. When we arrived home that night, she wrapped her finger with a proper bandage with two Popsicle sticks; one on each side, some disinfectant, and that was it. What she had done was cut her first finger on her left hand practically off, it was only holding by some skin, she would not go to a doctor, but you know, it healed up beautifully and has a scar to show for this experience. The next day we hired my brother Lorne who was home on leave from the army, and he could really handle this type of truck, paid him so much per thousand board feet for hauling and he really made some money while on leave, which I think was a week or ten days. We didn't quite have all the lumber hauled when he had to leave. The next driver I hired on the same terms and he rolled the truck and load over by trying to make the same time as Lorne but he didn't have the same experience with trucks and soon found this out, but we did get the lumber out. In those days you had to hire whoever was available as there was a real shortage of men during the war.

During the winter of 42-43 I talked to dad about buying the timber that he had on the farm as I had thought about putting a mill in there, as the railways were looking for someone to supply coal doors, it is actually a panel made out of rough lumber placed inside the door to stop the coal from rolling against the boxcar doors when they were hauling coal. Dad suggested we would work together on this project, and I know that I shouldn't have done that, as it turned out neither of us made any money. I should have let him go on his own, but hindsight is awfully good stuff.

In the spring of 1943 I went down to the farm, it had been vacant for a year or so, and started to prepare for a bush crew; also bought a good team of horses and hired a teamster who had worked for dad before, a real good man. I suggested to dad that we would move the house up on a ridge closer to our timber and besides there was a spring up there, a real good water supply. He thought I was nuts, well anyway he had to go to Port Arthur for a few days so we got busy and proceeded to move the house, it wasn't a difficult thing to do, anyway not for me. The house was a two storey frame about 20 feet by 40 feet.

I had procured a heavy jack somewhere, so started to jack the house up and slid round logs under the frame base, also had a set of block and tackle to pull the house ahead. With the team of horses and a post that is called a dead man set in the ground on an angle, we hitched the horses to the end of the block and tackle and away we went .We had logs laid crossways and as the house moved forward onto the logs in front we would pick up the logs from behind and place them in front of the house. Part of the way was quite level ground, then about the last 300 feet was a gentle slope, so much so that when the horses slackened off on their pull, the house wanted to roll back, so we had to place something behind the logs so they could not roll, this move was a good quarter mile and we were over half way when dad came back, we had hoped to have it completed before he arrived, and when he saw what we were doing, he said "you can't do this, you won't get it up that grade," well we said :just stand back and watch." The teamster assured dad that it was light work for the horses with the set-up we had. We moved it up to the top of the ridge and turned it twice before we had it in the right location just above the spring where the water was handy, and was on sand instead of where it had been in clay that stuck to everything every time it rained. Dad was quite pleased once it was there and said that is where it should always have been.

We then built a cookhouse and dining area for our crew also an ice house for storage of food. While cleaning up around the buildings also burning grass, a good building filled with machinery caught fire and burnt and dad was some upset, a real loss.

As usual there was a shortage of men, there were quite a few young Quebec Frenchmen available, they were the ones that didn't want to be soldiers and wanted to get in some bush camp out of sight. Occasionally you would hire one who was a real joker and we would get some laughs; our sawmill crew were mostly all Finns and we used to joke about having three crews, one working, one drunk, and one sobering up, as it turned out, that was what we had to put up with. All these men knew they could work any day they wanted, whether it was for us or someone else, so every other week the working crew went out for a drink, like it or not.

In this operation mother did the cooking, Merelie and my sister Marion, plus 2 or 3 other women nailed coal doors together, I hauled the doors to the railway and whatever had to be done and dad was foreman. This went on during the summer and fall of 1944. Our son Keith was around with grandma throughout the days, but he at 2 years old would try to get out in the bush if we didn't keep tabs on him.

Along in the fall, we had to hire one of the local fellows from the community to tail saw, as some of the others hadn't returned from their drunk. Oh yes while mentioning drunks, we always had to drive in to the city to pick them up, of course we were not the only one having this problem. There was one fellow who had a crew at the same hotel as our men, he was a big raw boned Swede about 6 feet 4 inches, this day he backed his one ton panel truck up to the door of the hotel and went with a baseball bat, he chased out a bunch of men and as they came out of the door he would tap the ones he wanted with the bat and in the van, and after he had his load, he turned and said "you want me to get you a crew too, dat's da vay I do it," but it was almost the truth because they never wanted to return to work if they had any money or credit. This particular hotel was owned by a big man known as Blind Swede. We used to hear some fantastic stories about him.

After hiring first one, then two local fellows to tail saw the portable mill we had, it turned out they were too lazy and we would have to go out in the evening and pile lumber, so Merelie said "fire these fellows and I'm going to tail saw." It was heavy work and we didn't want her doing that, besides we were expecting our second child that fall, but no way could anyone talk her out of it so she started tail sawing, kept up to the mill and did it for two weeks until one weekend she had cramps in her stomach and we had to rush to the hospital where our daughter was born. This was an experience, we didn't have a car at the time so had to get someone else to take Merelie in. As it turned out, it was a fellow who drove 30 miles an hour at top speed, his wife and Merelie were afraid the baby would be born before we could get there which was about 45 miles, and as they traveled they kept telling him to hurry. Our daughter was born just as we arrived in the hospital, a real cute baby girl who we named Faye Merelie, this completed a proud family. When I went back to camp the next morning and said we had a daughter, the crew wanted to know where Merelie was as they couldn't believe that she had a daughter, never showed any sign of being pregnant and had been working like a horse. Faye was born about two weeks earlier than expected on November 6th 1944, real healthy and you could see the mischief in her eyes.

It was much different than before Keith, our son was born, Merelie really showed for quite a while and her feet would swell. One time we went to a theatre to see a show, and while there Merelie's feet started to swell so she took her shoes off, and when it was time to go, she couldn't get her shoes on so I went out to the car and brought in a large pair of leather mitts which she had to wear out of the theatre.

We closed down shortly before Christmas and everyone left except Merelie and I and Keith and Faye, at Christmas some of the family were there. We had moved into the cook house and made it into a small bungalow for ourselves as we had most of the timber cut. We had a good Christmas, and after the holiday I decided to do some repair work on the mill, had to lay on my back in the cold and no doubt some snow, this I did for several days while doing the repairs, and after several days of this, one morning I could not get out of bed, in fact I couldn't move any part of my body except my head, I was completely paralyzed, was really painful when I tried to move. Merelie had to feed and shave me and everything, this went on for several months, and I assumed it was rheumatism as my back really hurt, a lot of men working in the bush had back problems and rheumatism. After some time, one day we had a couple of fellow who lived in our area come in to see me when they came back from the bush camp in the spring, I'm still in bed, so one of them, a Norwegian fellow said Merelie should put sulfur in wool socks and put them on my feet and I would be able to get up and walk, incidentally, these two men had been in the same camp where I had my back injured several years previous, and most likely this is from that injury. When we started thinking about it, I had been turned down by the air force, and then rejected from the army service because of my back, so up to this happening, I had always felt good, now my past was starting to catch up with my future. Well anyway we started to use the sulfur and my feet would perspire, in fact the socks would be wet in 20 to 30 minutes. I had never perspired before, but I was moving my toes and hands and showing some improvement. While I'm lying on my back, I started reminiscing of the past and maybe what I should have done instead of this or that. Late in the fall one of the owners of a manufacturing mill I had supplied with lumber called on me and wanted me to go and start buying for him, but I didn't think I should leave dad at this point so I said no, but now I know I should have taken that position, and I wouldn't be crippled up like this. I know now when I think back on it, it was a good opportunity to get my family away from all the hardships and work there was here, as I've said before hindsight is awfully good stuff if it worked. Eventually I was able to get out of bed with the use of sulfur, and immediately went to see our doctor. When I explained how I felt and my problem, he asked me what kind of ring I had on my finger, I said silver, he asked me why it was black, so I explained that I had used sulfur to be able to get mobile, he said "with your silver ring that color you have so much sulfur in your system, you are lucky you never caught a cold because it would have immediately turned into pneumonia and you would have died, so don't use any more sulfur." This doctor's brother was married to Merelie's cousin and was considered like one of the family; he told me there were some cures that were being tried but not much success. If we wanted to move to a warmer climate like Arizona he would pay for it If we didn't have the money. When I thanked him but said no, he then said "I'm going to send you to a friend of mine who is helping people with arthritis, which is no doubt what you have," as by this time he had an x-ray of my back and it showed a large area on the left side of my back and hip that had been bruised and there was coloring like blood clots showing, anyway I went to see his friend who used herbs for a cure, he gave me a quart sealer of this weed and said to start making tea out of it, drink a cup with each meal and any time you want a drink, have this, no alcoholic beverages, tea or coffee, or any type of caffeine, also no cured meats like salami or those types of meats. This weed won't cure you but it will keep you mobile and stop the pain. I said "what is it?" and he said it was yarrow, it grows everywhere and you can pick it, replenish your supply every summer. Well I started drinking this yarrow tea, it is bitter but I started feeling much better and continued until I was back to normal, then as he had said, it never cures the arthritis but keeps you mobile, so whenever I would feel the problem return, back to the yarrow and continued until I was in my late fifties.

The spring of 1945 we moved back to the city to our house on Martha Street and carried on with purchasing some lumber, also now looking around for older houses that we could purchase for remodeling; we found one on the corner of River and High streets in Port Arthur. There were a few old houses around that a handy person could buy and remodel and make a good profit at, so this was our plan, whenever there was any spare time. I always had business people who would loan me money if I needed it. The house at the corner of High and River streets was an old bungalow on a double lot; we had to do quite a lot of searching to find the registered owner who was an old lady in a retirement home in the city. The taxes were being paid by the daughter who had changed her name for some unknown reason and the house was rented out, it was quite a trial to finally get to the actual owner, and when I did, the Catholic home operators didn't want her to sell, as they no doubt felt they would eventually have it. I also had to locate some of the heirs, even her son who worked in the local post office and didn't even know where she was. I couldn't believe the situation, and she was a nice old lady and I went to see her several times as apparently she had none of her family calling on her, however no doubt after I bought her home, she had company as she had some money.

We moved into the house and started remodeling or improving it as we had time. Some of the repairs were a new roof, water lines, and some remodeling inside; refinish a beautiful hardwood floor in the living room which was quite large. One day while roofing, Merelie called me in for lunch and I left the ladder leaning up against the edge of the building or roof. After lunch Faye went outside to play before I was ready, well anyway that day my mother and dad decided to come and visit us. When she came to the door, she was really frightened, and said "do you know where Faye is?" we said "yes outside playing," she said "yes, up on the roof," said she was running back and forth up there and had come right over the edge to say hello to grandma and almost frightened her to death. There was one thing for certain, Faye had no fear of heights, and she wasn't 2 years old and would venture anywhere. Always had been a traveler or whatever from the time she could crawl. I had made a walker so the children wouldn't have to crawl and it would help them walk, Faye really made use of that, it sometimes took three people to keep track of her; she liked to tease Keith, her brother. We had an open bookcase and Merelie liked to keep the books lined up nice and straight, Keith liked to have them as his mother wanted them, so every once in a while Faye would scoot over to the books and run her fingers along them in one quick move and Keith would patiently straighten them all out and tell Faye she shouldn't do that. The two children were so much different, Keith was like me and Faye was like her mother, full of mischief. Some days Faye would get up in the morning and say "I'm going to be a good girl today" and she really was, but watch out the day she didn't tell you, she could find something to get into and it usually was something we had to laugh about.

At the front of our house was a stop for the street cars, so one day Faye decided to go for a ride on a street car, now remember Faye wasn't 2 yet and Keith not quite 4, she went to the stop and got on when some other people were boarding, took a seat and went along for the ride, now Keith who was supposed to be watching her, or let us know if Faye went out of the yard, saw her get on the street car but couldn't get there quick enough to catch, so he waited for the next car and got on, and hadn't said anything to his mother, these cars came around every half hour, the one that Faye was on was downtown when the one that Keith got on at our home; he thought he was going to catch Faye, unknown to us, Faye knew where her home was and had got off when it arrived back home. While all this was going on, no one was aware of the situation, anyway my mother boarded the street car downtown and saw Keith, asked where his mother or dad was and he said at home, "did anyone know he was on the street car?" and he told her about Faye and said he wanted to catch her before we found out that she was on a street car, he of course didn't know there was more than one street car, well my mother almost had a whatever, here are two children riding around the city on separate street cars and no one knows about it accept grandma now, she got off with Keith at our stop and came in the house and we are wondering where Keith was as Faye had got off at the right stop and was back home, the Lord was looking after our two little angels.

We used to visit mother quite often, she lived about 2 miles from us, however another time while we were living on River and High, Keith decided to walk down to see grandma, another day when we were both busy working on the house, he had walked to grandma and grandpa's house when some lady saw this little boy by himself and asked him where he lived, he told her, but he said "I'm going to see grandma and grandpa." "Do your parents know this?" and he said "no, grandma could phone when I get to grandmas." As it turned out grandma and grandpa were not at home, well this lady got all excited and she phoned the police, they came and picked him up, he told them who he was, his address and phone number and where his grandma and grandpa lived, so as no one was home, he brought him back to our house, which really surprised us, the police said "you really have a smart son, knew his address and phone number." Keith never worried about getting lost, even around the bush camps we had, he would go out on his own if we didn't watch him.

We had some interesting things happen while at this home. One time Faye took her little wagon, went across the street to a neighbor who had a beautiful pansy garden, she went into the center of the garden and filled her little wagon full of pansies, just the flowers and brought them back for her mother, when Merelie saw this, she sent Faye back to apologize to the neighbor for picking those flowers, but first, she had her wagon of flowers so unknown to Merelie, put the flowers all back in the garden and then apologized, to the neighbor's dismay, she had walked all over the flowers the second time, but Faye explained that she wanted some flowers for her mother, but wouldn't touch her garden again and was sorry. Our neighbor said she couldn't believe what was happening to her pansies when Faye came back the second time and returned them. Another time Faye had some fun to Merelie's dismay, the city had just sprayed the streets with a tar substance, and Faye had put on a long slip and a pair of Merelie's shoes and went out walking up and down the street, and of course was covered with tar, I think this time she was trying to be a lady.

In order to keep the children home, we put a fence around the property, it was a large double lot in grass and some of our neighbors used to walk across the lot next to the street. Well shortly after we put up the fence, one of the neighbors said to the children "I see your parents put up a fence to keep you home." Faye who could talk real good when she was one year also walked at nine months, said "no not to keep us in but stop the neighbors from walking across our lawn" boy, how to keep in good with the neighbors, one of them told us about it later.

This home had some good memories and one that really upset us at the time, however first the good memories. We had a few good house parties; this home had a large living room with a really good hardwood floor, good for dancing. One time a salesman called on us selling Tupperware pots and pans, well he wanted to organize a dinner party where he cooked the whole meal and endeavored to sell as many pots and pans as he could. As it turned out, that day we had some company come in from out of town who also knew quite a few of our friends, so Merelie got the salesman all geared up to put on a dinner that day, she invited in several more couples, in fact there were twelve couples that sat down to eat that afternoon, of course in those days there had to be some alcoholic beverages to go with the dinner, well anyway it turned into a real party, the salesman sold 12 sets of pots and pans and he went home feeling no pain and it took him a day or two to get over that. When all the dust settled, no one cancelled their purchases and that salesman said he had never done anything like that before and thought we should do it again.

After we had this home remodeled, we moved out to another adventure and rented the house out. The couple that rented were middle aged, recently out to Canada from Britain, seemed to be a nice sensible couple. We lived out of town and would come in each month to pick up the rent. After they had been there several months, we came in one day and the tenant was all excited and wanted us to see how he had been painting; it was about the same time that some company had come up with paint that you could paint on wood and mottle or grain it like hardwood floors. What do you know, he had painted this beautiful hardwood floor, wow, I could see red, no doubt he had done a job painting, gave the tenants their notice to move right there, he wouldn't try that on a hardwood floor again I bet. Anyway we were so peeved we sold the house, made some money but maybe should have left it rented, he had done it and was real pleased and may have stayed there forever, who knows.

We had moved back to Dorion where we had purchased a stand of timber and a camp on the Brunner Road area, this timber my dad had owned, I believe it had been homesteaded by his brother in 1903, and he had died before any improvements had been made, although part of the timber had been cut previous to our purchase. We moved into the camp, which consisted of one building, bunkhouse in one end to accommodate about 16 men, and in the other end a kitchen and dining area, with one room off the dining room. We also built an office separate from this building which we would also use for our sleeping areas as we had our children Faye and Keith and it would be quieter for them in the evening.

When we went down to Dorion to this camp it was late in the fall, we planned on renting a sawmill with crew and also cut pulpwood as we had a contract to sell pulpwood to the Red Rock Mill by now. We had to wait until there was sufficient frost to be able to haul out the lumber and pulp. Merelie and I moved into a room of the dining room in the large building and had grandma looking after Faye and Keith until we could get organized. We had a large tin heater stove for heating this building with a 7 inch stove pipe that went straight up through the roof. I had put on an extension on the stove pipe to take the cold air off the floor, this extension protruded out behind the stove then back down to about one foot above the floor, it worked really well. We could keep the floor real warm with this suction from the stove pipe, the hotter the stove would get, the more air would be drawn up the pipe. Well one evening dad had come to stay with us over night and we were sitting around discussing whatever and as it was a cold night, keep putting wood in the stove, and these tin heater stoves had a tendency because they were light weight, of jumping up and down if the stove got too hot, that is what happened this time and the stove all of a sudden turned red, well we were a little concerned, as with the stove jumping around, it could knock the stove pipe down and we would have the camp burn down. The first thing I thought of was the pail of water, instead of pouring in the stove which may have been a little dangerous, I placed the pail under the pipe that reached down to the floor, and wow, things really happened, there was enough draw that it sucked the pail dry, the stove went black and as the water went up the pipe so did all the wood and ashes out of the pipe, which no doubt was quite a sight from outside. At the time I placed the water under the pipe, a couple who were friends of ours were on their way to the camp door for a visit. When all this happened, they said there had been a red flame coming out of the pipe and all of a sudden there was a big cloud of steam and out of the cloud of steam came tumbling pieces of wood which bounced off the roof, they couldn't believe their eyes, and waited a few minutes before knocking to make sure they were safe, while inside we had experienced the sound of the wood going up the pipe and then bouncing back on the roof, rather startling experience to say the least, but the consolation was we stopped the possibility of burning the camp, and just had to start the fire over again but from then on we would be much more careful, had a good laugh about our experience and also our friends.

That same fall while preparing for the winter operation we also had a large cook stove which I would light in the morning and then jump back in bed and wait for the camp to warm up. We had a very smart black and white Cocker Spaniel with us there, and he was always with us in the camp at night, he had a sleeping area in the kitchen. One morning I heard Klipper growling and woof and all of a sudden he came running into our bedroom and jumped up on the side of the bed and then back in the kitchen and then right back and this time jumped right up on the bed, he never did this before so I ran out to see what was up, here was Klipper chomping on a piece of wood that had fallen out of the stove and was burning the floor, I took some water and put it out and then gave Klipper a real good petting, it was quite possible he had saved our lives, had I fallen asleep, which did happen once in a while. After this happened Klipper always watched anyone who smoked, as their cigarette was fire and he would bite the cigarette right out of their hand if he could get close enough to it, and would drop the cigarette on the floor or ground and jump on it with all feet until there was no fire.

We finally got everything organized, had a fellow bring in a sawmill with his crew and also our men for cutting logs and pulp, they were all young men who would work anywhere they could find work as now it was different from during the war when there was very few men available. We used to laugh about some of the nicknames that some of these fellows had, there was "Molly" a big rugged fellow, and another fellow "Shirley" and another tall lanky fellow "Mabel." Where they got these nicknames I'm not aware of. Mabel had come from an orphanage in St. Vital near Winnipeg, had joined the Merchant Marines at about 15, been torpedoed a couple of times, had a nervous breakdown and was discharged from the navy, he was now17 when he came to work for us. The employment agency asked if we would take him as we had a small crew and could get the men in the bunkhouse to understand his problem, which they all did, and during his sleep he would dream about being torpedoed and start yelling, so someone would have to talk to him and tell him he was okay and where he was and he would go back to sleep, it didn't take him long to become normal during his sleeping hours and was certainly a good worker and nice fellow. He would often come and visit with Merelie and me in the evening and he liked the children and they like him, Keith and Faye were now staying with us. This lad became quite a successful businessman in time. When I met him several years later in the town of Geraldton, he was married, owned a fleet of buses and small trucks which he had working for one of the timber companies, also owned the Chrysler dealership in that town. We were real pleased to see how well he had done. Another fellow who worked for us that winter went on to become the chief forester for the forestry dept. at Port Arthur, his name was Wally Jarvis. Merelie did all the cooking for this crew, I believe we had 16 employees, It was certainly a full time job, there was breakfast at six, then coffee and cake for anyone that happened to be at camp at 9:30 then a full meal at noon, for those that were out in the bush all day would take their lunches, which made a lot of work, then at 3 p.m., coffee and cakes etc. and then a full meal at supper time, and we did have good meals in order to keep our men happy and working. We did have a couple of duds but weeded them out, in fact Merelie refused to feed one fellow who wouldn't wash and the men complained about his odor. When I went out to scale his logs and pulp he owed us money, so besides being dirty, he was lazy. Each man paid so much for board and there was this time and the stove all of a sudden turned red, well we were a little concerned, as with the stove jumping around, it could knock the stove pipe down and we would have the camp burn down. The first thing I thought of was the pail of water, instead of pouring in the stove which may have been a little dangerous, I placed the pail under the pipe that reached down to the floor, and wow, things really happened, there was enough draw that it sucked the pail dry, the stove went black and as the water went up the pipe so did all the wood and ashes out of the pipe, which no doubt was quite a sight from outside. At the time I placed the water under the pipe, a couple who were friends of ours were on their way to the camp door for a visit. When all this happened, they said there had been a red flame coming out of the pipe and all of a sudden there was a big cloud of steam and out of the cloud of steam came tumbling pieces of wood which bounced off the roof, they couldn't believe their eyes, and waited a few minutes before knocking to make sure they were safe, while inside we had experienced the sound of the wood going up the pipe and then bouncing back on the roof, rather startling experience to say the least, but the consolation was we stopped the possibility of burning the camp, and just had to start the fire over again but from then on we would be much more careful, had a good laugh about our experience and also our friends.

That same fall while preparing for the winter operation we also had a large cook stove which I would light in the morning and then jump back in bed and wait for the camp to warm up. We had a very smart black and white Cocker Spaniel with us there, and he was always with us in the camp at night, he had a sleeping area in the kitchen. One morning I heard Klipper growling and woof and all of a sudden he came running into our bedroom and jumped up on the side of the bed and then back in the kitchen and then right back and this time jumped right up on the bed, he never did this before so I ran out to see what was up, here was Klipper chomping on a piece of wood that had fallen out of the stove and was burning the floor, I took some water and put it out and then gave Klipper a real good petting, it was quite possible he had saved our lives, had I fallen asleep, which did happen once in a while. After this happened Klipper always watched anyone who smoked, as their cigarette was fire and he would bite the cigarette right out of their hand if he could get close enough to it, and would drop the cigarette on the floor or ground and jump on it with all feet until there was no fire.

We finally got everything organized, had a fellow bring in a sawmill with his crew and also our men for cutting logs and pulp, they were all young men who would work anywhere they could find work as now it was different from during the war when there was very few men available. We used to laugh about some of the nicknames that some of these fellows had, there was "Molly" a big rugged fellow, and another fellow "Shirley" and another tall lanky fellow "Mabel." Where they got these nicknames I'm not aware of. Mabel had come from an orphanage in St. Vital near Winnipeg, had joined the Merchant Marines at about 15, been torpedoed a couple of times, had a nervous breakdown and was discharged from the navy, he was now17 when he came to work for us. The employment agency asked if we would take him as we had a small crew and could get the men in the bunkhouse to understand his problem, which they all did, and during his sleep he would dream about being torpedoed and start yelling, so someone would have to talk to him and tell him he was okay and where he was and he would go back to sleep, it didn't take him long to become normal during his sleeping hours and was certainly a good worker and nice fellow. He would often come and visit with Merelie and me in the evening and he liked the children and they like him, Keith and Faye were now staying with us. This lad became quite a successful businessman in time. When I met him several years later in the town of Geraldton, he was married, owned a fleet of buses and small trucks which he had working for one of the timber companies, also owned the Chrysler dealership in that town. We were real pleased to see how well he had done. Another fellow who worked for us that winter went on to become the chief forester for the forestry dept. at Port Arthur, his name was Wally Jarvis. Merelie did all the cooking for this crew, I believe we had 16 employees, It was certainly a full time job, there was breakfast at six, then coffee and cake for anyone that happened to be at camp at 9:30 then a full meal at noon, for those that were out in the bush all day would take their lunches, which made a lot of work, then at 3 p.m., coffee and cakes etc. and then a full meal at supper time, and we did have good meals in order to keep our men happy and working. We did have a couple of duds but weeded them out, in fact Merelie refused to feed one fellow who wouldn't wash and the men complained about his odor. When I went out to scale his logs and pulp he owed us money, so besides being dirty, he was lazy. Each man paid so much for board and there was also the van with items such as chocolate bars, tobacco, and some tools which they would charge up.

That bush operation turned out to be very unprofitable; never in that area was there a mild winter, the winter of 1947 and 48 there was very little frost and we had to haul our lumber and pulp over a swamp that would not freeze so although we did get some out, we had to stockpile it until the next winter, or build a road out over a mountainous area, we decided to stockpile, however we now know we should have closed up for another year, as it turned out to be a disaster. In the very early spring I had to go into the hospital for an appendix operation, and had just come back to camp when we had to rush Faye into the hospital and she had appendicitis also, very seldom a 4 year old child has appendicitis, but there was the case. I had planned on having the timber that was cut in the bush insured until we could haul it, however I hadn't and some way or another a forest fire got started and burnt everything except the camp and the mill. We had all our savings go up in smoke and came away from there broke, and to make matters worse, one box car load of lumber we shipped, we never did get paid for through a crooked lawyer. This was close to two thousand dollars, a lot of money to us at the time. During the early spring it was a nice warm day and most of the men including myself were sitting outside on some logs that were between the bunkhouse and the toilet and behind it we had been throwing empty cans and various types of garbage that didn't smell or bring flies, anyway a black bear who had probably just came out of his den was out looking around our garbage behind the toilet, we had been watching him; out of the house came Mabel who had apparently been sleeping, and headed for the toilet. Mabel took long steps and he had his head down and as he approached the toilet door the bear came around from behind the toilet, and Mabel and the bear met head on, they were both startled, the bear stood up on his hind legs and turned and left and Mabel let out a holler and turned and went back toward the bunkhouse as fast as he could, it was really a funny thing to see as each went in opposite directions, by the time Mabel was back at the bunkhouse the bear was long gone into the bush.

One morning one of our cutters Sye from Nova Scotia went out to cut. (We had nicknamed him Sye as this is what he would say, whenever he was going to say anything he would say "sye.") In about 10 minutes I saw him coming back, I asked him what had happened, had he forgotten something? He said "no, I met a "bar," I said hello Mr. "Bar" and he didn't answer so I came back until he went somewhere else, sye that "bar" wasn't very friendly.

Besides the timber camp on the Brunner Road in Dorion, we also owned a small farm 160 acres with a good hewed log house, about 35 acres cleared, also an 80 acre parcel with good growth of Christmas trees that cornered the farm. About the same time the bush burnt, someone threw a cigarette down along by our farm and burnt the log house and the stand of Christmas trees, which would certainly have helped us financially the next winter. It seemed as though someone was out to get us out of the area, however we did stay around the area for another couple of years.

During the late 1940's our little family had some detrimental times, with my wife Merelie and my son 4 years and a daughter 2 years our financial position almost became nil.

We had been in the lumber and pulp business in a small way in Northwestern Ontario for several years. I bought a stand of timber and camp with my cash flow and operated it during the winter of '47, but there was little frost and we had to stockpile a lot of lumber and pulp which we would haul in the summer.

As I had been raised on a mixed farm in the same area, mixed meaning milk cows, pigs, and market gardening, all done with horses during the great depression, I knew how to work and my wife was a good helper, this meant only one thing, get busy and try again to succeed. We analyzed our position, no income, a small farm, 160 acres with 35 acres cleared in hay, a shack and barn but determination to succeed.

At that time there were Dutch immigrants moving into the area wanting to buy farms; also there was a new dairy plant which was giving out milk quotas to improve their milk supply. I immediately proceeded to find some milk cows; we would build up a small dairy farm and sell it as we had other plans for our future. I found a dairy farmer who was retiring and arranged to buy 16 cows, to be paid for out of the milk the cows would produce, and if I failed, the cows would go back to him.

We were presently living at my dad's farm which was about a mile away, we had to move on our own farm; eventually used one of the lunch shacks that had not burnt and was used in our woods operation, these shacks were 10x16 feet on skids so they could be hauled around to where our bush crew worked, this was very temporary, however it was our home for almost 3 years.

Now we proceeded to build a small barn for 16 cows which arrived as soon as it was suitable to use. Now we were in to dairy farming. We had to build some fences to keep the cattle out of the hay fields and pastured the cattle in the bush; fortunately there was lots of grass. When we would bring the cows in for milking, Merelie would clean or wash their udders and tie their tails up on a wire strung along behind the cows and I would milk each cow by hand, if I remember correctly we sent out eight 8 gallon milk cans daily which of course gave us a good income in those days and certainly was appreciated after our previous set back. We didn't have much money to spare after making a payment towards the cows and purchasing grain for them to keep the milk production up, however we were succeeding.

We did purchase a brood sow that had 12 piglets shortly after we bought her, these were for any milk we had over, especially during spring and summer months, the pigs drank the surplus milk and we would sell the pork in the fall. We went through our first year with flying colors, also bought a pony and cart from our neighbor for the children to enjoy and it also came in handy for running errands as we only had an old Chrysler car that had to be parked at the top of the hill to start it as low gear was stripped, the finance company had seized our van after the fire.

The second year we bought a used homemade tractor, an awkward thing but it worked. We also bought a used homemade buck rake which saved a lot of labor while haying. We built a hay barn on the end of the cow barn and installed a hay fork in the roof to pick up the hay off the buck rake. For your information a buck rake has about eight 8x10 foot prongs which pick up the loose hay, ours was built on the rear end of an old truck cut down for this purpose, the hay is raked into windrows, then you back the buck rake into the windrows until the rake is full then drive to the barn where the hay fork in the barn picks it up and dumps the hay in the barn.

We went through our second year quite well. We now almost had the cattle paid for, had a few acres of grain in our second year which meant we didn't have to purchase that, and were now thinking about building a house and maybe buying a used truck to help us around the farm. Would you believe it, the dairy where we shipped our milk started getting behind in payments for our milk, this was along about April of 1948, finally with five months behind in payments for milk the dairy closed, the owner said due to competition but had promised to pay for our milk which never materialized. Now we have no place to ship milk, we did try shipping cream, however the price was so low that it didn't pay, we had more pigs which I believe got up to 48 that fall.

One morning along later in the summer we were going out to milk and our daughter ran on ahead, and she saw a black bear looking at the pigs, bears really like pork, she ran back and told me, so I took the rifle as it may have to be destroyed, however it had left fortunately and we never did see one again.

We started to sell the cows, as with no milk quota you couldn't make a living, and by late fall we sold them all except one Jersey which we kept to have our own milk and butter supply.

One day when Merelie had the old car in Port Arthur, the forward gear all stripped, she could only go in reverse, so she backed into a parking lot and left it there, went to one of the dealers and bought a used Fargo ½ ton truck, the body needed repairs but the motor and main parts were all good. We later fixed it up with a flat deck and used it for several years.

That same fall there was an early snow fall and we lost our grain crop the day before it was to be combined, the pigs salvaged part of it but that didn't help our finances much. We finally sold all the pigs except our original brood sow, and in the end had to butcher her which we used for our own meat. We never wasted anything that was edible, but there were times we wondered about fat pork.

Our son started school in September 1948, he had to walk about half mile which was down a steep hill, cross a bridge over a good sized creek, then catch a school van that took him another 3 miles to school and of course return after 4 in the afternoon. We were concerned for him being alone at six years of age, however most children growing up in the country are quite accustomed to this, and he did very well. Our daughter was coming 6 years and started school in 1950.

Summarizing our position, we were still in the 10x16 foot shack which we call our home, have a few dollars from the sale of the cows and pigs and have one milk cow, a few chickens, and a pony and a cart. No income, looks as though it will be a long cold winter.

We still wanted to sell the farm as it is and move into the city where I could get work as a carpenter as it is my trade. I approached the farm board at Fort William and told them my sad tale and said I would sell for whatever they could get, which later materialized and we sold the farm in early 1951 for $780.00 after all the work we had put into it, but better that than nothing. We also owned the 80 acres where the Christmas trees were burnt, also the timberland that burnt we could not or did not find buyers for these properties and they were eventually sold at a tax sale.

Keith who was now seven had been in school a year and a half was small for his age but quite sturdy surprised us one evening after school. There had no doubt been some discussion about Christmas approaching, how we had to watch our pennies if we were going to move into the city in the spring where I could work at carpenter work and have a steady income again. One morning when Keith went to school shortly before Christmas, he took an axe with him which we were not aware of, hid it near the creek at the bridge until he came home. Keith was late coming home this day, we were concerned, and waited a while before starting to look for him. Sometimes the van would be late but was becoming serious and it was getting dark. We decided it was time to go and look for him, we just started out down the steep hill and around the first bend, here comes Keith dragging a Christmas tree three times his size and carrying the axe. We couldn't get over the size of the tree and the size of Keith, he had quite a time getting this tree which was a real nice one. I asked where he had got the tree from, and he told me, and it was hard to believe. I went to see where the following day as it seemed impossible. He had climbed up the bank; the road had been built down a ravine to the creek, with a ravine on one side and a steep bank on the other; anyway it was a real steep bank where several feet of snow had drifted in, he had to dig a hole around the tree to get to the bottom to chop it off, he could have been buried in that snow, however I assume he was light enough that the packed snow carried his weight. The tree was 75 to 100 feet above the road, real steep bank, so after cutting it off, get it out of the hole, and I presume it rolled down onto the road where he then proceeded to drag the tree home, about 150 yards or more. I questioned him about why he had done this and didn't he realized we were worried and that he could have suffocated in that deep snow if it didn't carry him, his answer was, he thought he should help by getting a

Christmas tree as his mother and I had been discussing our finances and didn't have an income, and he assumed that maybe we wouldn't have a tree this Christmas. We could not get over the effort and determination he had to get a tree like this, and he said he had it picked out before the snow came and had thought he and I would get it before Christmas arrived, which no doubt we would have. As it turned out we had a lovely Christmas with our little family with a lovely tree, no turkey but a large chicken roasted with all the trimmings, the gifts were not plentiful, I believe one each in our 10x16 shack, warm and cozy, never the less we thanked the Lord for our blessings, we were all healthy, had each other to love, a solid roof over our heads, our little home was filled with the aromatic fragrance of the tree and a delicious dinner. This is one Christmas that has been remembered and been outstanding because of our son's determination to have a tree.

We have to tell about this as it was quite a joke about Keith and dad. One day in school Keith's teacher was asking each pupil what their ancestry was, or did they know their ancestry, so when it came to Keith he said he wasn't sure, however he knew his grandfather was a native Indian, so the teacher asked how he had arrived at that conclusion, as Keith was real blonde, and he said "my grandfather talks the Indian language" which was true, he spoke several dialects and used to teach Keith some of the language, the same as he had taught me. As we were well known in the community and most people knew where he and his sister had come from, so the teacher made it a point to call on us that evening so that Keith would get his ancestry straightened out. However as far as his grandfather was concerned he respected the Indians, and said he thought they looked after their families much better than the white men, that was in the early 1900's. He had worked with a lot of the natives on the survey and laying of steel through northwestern Ontario for the C.N. Railway.

When Faye started this same school, she thought she should be the teacher's helper, and sometimes as the teacher was walking around the classroom Faye would follow her or sometimes just not settle down as was expected in those days. One time when the teacher was tired of trying to get Faye to do as the rest of the class, the teacher sent Faye out in the hall to think about what she should do. Well while Faye was out there she decided to sing "I've got a lovely bunch of coconuts" at the top of her voice and this distracted all the classes, so the principal came and had a talk to her about keeping quiet, so she said "okay Mr. Clackworthy" which was his name, however after a while she started singing again, anyway the outcome resulted in Faye's teacher and the principal calling on Merelie and I that evening and suggested we maybe should keep Faye home until she was more mature as they put it. Merelie said "do you think she isn't intelligent enough? They both said "oh no, she is very smart, and a loving girl, but the problem is she has us both wrapped around her little finger." Merelie said she needs discipline if she doesn't do as she should in school, she would behave with me and they said "oh no, we couldn't do that , she is such a sweet child, however she does disrupt the class and if you don't want to keep her home, we shall talk to the inspector on his next call." Merelie said "don't you think she is a little young to be expelled, of course if you wish to talk to the inspector you do that and tell him to come around and see us, we know the family really well as his oldest daughter was my girl friend and our bridesmaid when we were married." Well they soon changed their mind and Faye went into the principal's office for a while until she settled down. She gave Keith a hard time for a while, said it didn't take her several years to get into the higher grades. Keith wanted to stop going to school as he was embarrassed by Faye's behavior.

We had purchased a Shetland pony and cart for the children while farming, as previously mentioned. The pony would come to Faye any time and follow her around, even though she would pull tricks on Sandy, like at the electric fence she would touch the pony and then reach over and touch the fence and Sandy would get the shock, of course she would do this to anyone that wasn't aware of what would happen, she shocked her grandmother several times and though it was a real joke. One day I saw Faye hanging onto the fence with both hands, so I said "doesn't that bother you? And she said "no, only it feels funny." As to Keith and the pony, he could call and call and Sandy wouldn't come to him, he would have to go out and catch him if he was going to use him. Sandy and Keith got along really well when he was using him but Sandy liked Faye.

As mentioned before, we sold the farm or should I say gave it away, but better that than nothing. At one time I wanted to purchase the farm when I had the money to do it from mother and dad and also purchase some adjoining land in that area, it would have made a good ranch with the Coldwater River on one border, the Spring Creek on the two other borders and the mountain on the back side. With some additional fencing, it would have been approximately 900 acres where we could run cattle, and at the time my dad and uncle thought I was crazy, however my cousin did it some years later and they are into the second generation with beef cattle in the area and doing very well at it, so it wasn't such a crazy idea, the winters are no different from the caribou area in B.C. where ranching has been carried on for years.

We took over a service station at the corner of River and Balsam in Port Arthur, changed the living area into living quarters and started selling gasoline, and also had two service bays where we had a mechanic working for us. Living right at the station Merelie could look after the garage during the day and this left me free to work at carpentry for the summer and in the winter I went out in the bush and was foreman in the bush camp for the mayor of the city, Charlie Cox. This was in Lac Suel west of Port Arthur. I was there until spring when the ice was starting to melt and wasn't safe to travel over. When we started to close the camp down, there was some supplies that had to be moved out and although we had been warned about the ice on this lake, one of the teamsters decided to cross the lake rather than go around it which was a lot further, however he lost his team of horses and the sleigh through the ice. The next day Mr. Cox, the owner came in with an enclosed snowmobile and took the cook and clerk out, also wanted me to go, but I said "no I'll walk around the road" which was a full day but I didn't trust the ice as that was the owner's intent to cross the lake, they got out about half way when one side of the snowmobile went through, fortunately for them the left side went down, they were able to get out where the door was. They lost the snowmobile and had to walk the rest of the way on the ice which was about 12 miles. The ice on the lake was not like ice in most areas which was a larger grain or appears that way. On Lac Seul the ice would give way and a piece would drop straight down without warning, most ice would crack and be noisy but not this lake. I was glad I walked out and never did go back to that area.

In the spring of 1951 we bought a house on Hull Ave. that had been burnt in the inside. This one needed quite a remodeling job. That was one of the dirtiest projects we ever undertook. We put a cement block basement under the house, completely remodeled the main floor, also put in an oil furnace. We also made a suite in the basement, and prior to doing the remodeling we built nice garage and workshop.

One day after we had moved into the house a fellow came along and said "I'm glad you built me nice garage." So I asked him what he meant, well he said "I have two lots next to your property and your garage is on one of my lots." I said "Are you sure of this?" He was quite definite about it, but the only thing he didn't know was where his corner post was and I did. We were 3 feet inside our line with the garage which was legal in that area, so anyway in the course of the conversation, I said "if you are so sure, how about I'll buy your two lots, as I then could build another house next door which would be handy for me to work? He said "no I need that garage if I build a house here," so I said "suit yourself, go and have your lots surveyed and we will see about the garage after you survey." Well he had it surveyed, and if he hadn't got nasty about the lot lines I could have shown him the survey post which was at the corner of our driveway under about 3 inches of gravel, however the end result was he found out where the line was, he was so peeved he wouldn't sell the lots to me anyway.

When we moved into the house, we gave up the service station which we had been renting, however we had been letting some of our customers who were steady and they usually paid up every two weeks, however there were a few that didn't pay their full accounts, so when they heard we were leaving, no doubt started dealing somewhere else where they could charge; there was about a thousand dollars outstanding when we left the service station. It was an Imperial Oil outlet. It took me almost a year to collect those delinquent accounts. I remember one I called on one day, he owed over a hundred dollars, so when I knocked on the door the fellow who owed the money answered the door, you should have seen his face, he said " you know I was going to come up to your station this week to pay you." Well when I had the money in my hand I said "why didn't you come back before we left, we haven't been there for a year"? We had good neighbors there also, but usually always had good neighbors. Mother and dad moved into the basement suite and were quite happy there.

Prior to this I built a house for mother and dad on Edith Ave. with an above ground basement, in those days when I built, I did the wiring and plumbing as well as the carpentry, but this work had to be inspected but I never had any difficulty. On this particular house another small contractor had been talking to the plumbing inspector about the work I was doing, so the inspector said "don't worry if the work isn't done right I'll catch him " however he told me about it and when the inspector came in, he really looked over all the work thoroughly, passed it and left. He never checked to see if the water was on, so as soon as he left I turned the water on, and what do you know, I had forgotten to solder the last joint at the top of the hot water tank and water was spraying everywhere, boy would I have been in trouble if the inspector had said to turn the water on. Mother and dad sold that house after a year or so as I believe the stairs were bothering them, and in our apartment there were only 2 to 3 steps which was much better and mother had company when dad was away cooking, I think in 1953.

Along in late summer there was a position I heard about that was with North Star Oil Company who had their head office and refinery in Winnipeg. I contacted them, had an interview and was hired if I wanted the position, it was quite interesting, and the company was expanding. I would be in charge of maintenance and construction of new service stations or enlarging the present ones. My territory would be from the Manitoba border to as far east as they went which was about 200 miles east of Port Arthur, they would supply a vehicle and the salary was quite good plus traveling expenses. The only thing I was not too happy about was occasionally I would have to repair pump heads out on the garage sites and sometimes it could be outside in the winter which wasn't the warmest thoughts, they also supplied good warm clothes and a good parka, so I accepted the position. Then I had to go to Winnipeg for six weeks to work with their employee of the same position in that area. We decided this would be a good time for Merelie to have a break, so grandma had Faye and Keith while we were away.

When we arrived in Winnipeg we started looking for a small apartment as the company paid for my room and board, so would use that towards both of our living expenses while there. We were real fortunate and found a place the first day in a large house right across the street from the parliament buildings from a real nice couple who were retired wheat farmers from Saskatchewan. Roy McLeod was the wheat king of the world when he retired; it is quite interesting how he started and his accomplishment. When he was a young man during the depression in the 1930's he came out of school and as there was nothing to stay home for, his parents had a grain farm and couldn't sell any grain so he decided to do some traveling to see if he could find some work this was of course by freight train. He traveled across Canada then across the U.S.A., finally in California he had no money, nothing to eat, and decided if he ever got back to Saskatchewan he would grow a patch of potatoes and at least have something to eat. He finally made it back there still hungry, that was in the spring of 1937. He went to talk to one of his uncles about a patch of ground to plant some potatoes; his parents had lost or given up their farm, his uncle said he would give him a hundred and sixty acres to plant some potatoes, and when he had some potatoes planted, he also gave him some seed wheat to plant and said at least it will keep the soil from blowing away if it grows. He said the potatoes grew and so did the wheat, he had a real good crop and said he sold some of the wheat and kept some for seed and also had lots of potatoes to eat. He then arranged to get some land, planted that, had another good crop and sold it, from then on he kept enlarging his acreage until he was producing the largest acreage of wheat in the world owned by one man. By this time he owned most of the property around the town of Kamsack, plus a lot of the business in town, he had never had a crop failure in fifteen years, he said that was unheard of on the prairies, so he decided to retire because he had so much farm land that if there ever was a failure now, he could go broke, so started selling the land and businesses he now owned, he was now married and had a teenage family, so he and Mrs. McLeod went to Winnipeg and found a nice home, which is where we rented a basement suite. It was more like moving in with the family because a good part of the six weeks we had our meals with them. It was quite odd how we met Roy, we were walking along the street, and here was a fellow pitching horseshoes by himself. We stopped and watched him for a few minutes, he was quite good, when he saw us watching him he stopped and said to me "do you pitch horseshoes? And said I had done it a few times, so he said "come in here and have a game with me," we stopped and played part of a game and he asked where we were from and what were we doing, so we told him we were looking for an apartment, so he said "don't look any further if you only need a place for six weeks we have something I think you could use, but first come in and meet Mrs. McLeod," which we did, and everything was arranged right there, and said "now when you have some spare time you can play horseshoes with me." He was a young man to be retired, but was enjoying himself. He told me that when he retired his brother was the wheat king of the world. We used to go to church with them and they were real good company, she and Merelie got along really well, they could see the funny side of everything.

After six weeks of learning the way that North Star Oil operated I was prepared to head out on my own again, had a really good supervisor to deal with in Winnipeg, which I could not have said the same for the manager they had in Port Arthur, although he had nothing to do in my department, he couldn't mind his own business. He had been in the Imperial Army in India prior to coming to Canada; he was a real Imperialist Englishman and expected everyone to salute him every time you saw him. One time shortly after taking over my position at the Lakehead he called me into his office and asked me about associating with the men who worked for my department and said I shouldn't lower myself to their status as I was much above them. I informed him he was now in Canada and that we could get much more work done if the employees liked their boss than if they were treated as he would have them.

I had some interesting experiences while with North Star Oil, I built some new service stations or should say supervised the construction, also enlarged a number of stations, did some service work on pump heads. One time I'll never forget was during the winter at Red Lake at a mine there, they had an old style hand pump with the large top that held the gasoline which you pumped in before filling the vehicle; it had stopped working and needed some repairs, so I went to see what I could do with it. The evening I booked in to the hotel was real cold, the next morning I ventured out to repair this pump, it was up on a small hill setting out in the open where the wind could really blow and it was that morning which I didn't appreciate, anyway I dismantled the pump and repaired it, then put it back together and was back in the hotel before noon. The hotel clerk said "were you out for a walk"? and I told him what I had been doing; he said "do you know it was 50 below at 8am? I told him I knew it was cold repairing that pump, but that was my worst experience while with North Star.

The manager for the Lakehead division was not very popular anywhere in the area and of course kept getting in my affairs. One time a garage owner literally picked him up and threw him out of the garage. We almost lost a dealer there, if it hadn't been for me, who arrived shortly after. Due to this and other instances I talked to my supervisor in Winnipeg, I told him about what was going on and suggested they replace this fellow; he said he would discuss the matter with his management. He called me back and said that they wanted me to stay with the company and offered to let me buy shares in the company, and the manager who was a misfit in the North Star Oil was a major shareholder and they couldn't discharge him, but if we would tolerate him for another year or so they would move him out in the prairies where they were expanding. The offer of shares was a real good deal; you accept the shares in blocks of 500 at a dollar a share which was a good investment, the payment of shares was made by small deductions from your pay. They offered me several blocks, and if an employee did leave the company he could retain the shares by paying whatever balance he owed or they could buy them back at the going rate. As it turned out I didn't buy any shares as I knew I wouldn't put up with their manager for another year or so and planned on leaving the company in the spring of 1954. I should have bought the shares as several years later Shell Oil bought North Star Oil for $17.00 per share; could have made some money there but hindsight is not very profitable. I may have stayed with that position if it hadn't been for the division manager, I felt life was too short to put up with a person like that, I was my own boss and it was interesting with the expansion that North Star had going. I left them in the spring, was going to do some house building. One evening while Merelie and I were at the theatre I was talking to the theatre manager and he was looking for someone to do some maintenance on two theatres in Port Arthur, so I had a talk to him about the work; they paid $50.00 a week cash for part time work, carpentry repairs and painting or repairing whatever was needed, so the end result was I took it as I had quite a lot of spare time to do something else. Shortly after starting with the Famous Players Theatres, they wanted me to also be a doorman in the evenings 6pm to 11pm, so I took that on also, then occasionally there would be some type of stage show come to the city and they would have those on Sundays which I also looked after it became a full time position, however it had drawbacks, to start with those stage shows that traveled from place to place the makeup clothes that were used would stink, sometimes had to handle their trunks. I was concerned maybe getting some disease and bringing it home; six evenings a week I was doorman, had no time to spend with the family, and worst of all, the staff in those theatres were real party people, as soon as the theatre closed there was a party at someone's house, and of course we had to have parties also. We tried to avoid them sometimes.

One evening Merelie and I came home and here was a party in full swing, someone met us at the door and said "we didn't think you would mind having a party so made ourselves at home," well that was the end of parties and shortly after that I left the theatre and took another position selling for an American company by the name of Huntington Laboratories.

Merelie's dad had sold for the company and had done quite well, and as it turned out they let out a franchise for a territory from the head of the Great Lakes to the west coast. A man by the name of Wilf Graham, who was a commercial traveler, got the franchise and I worked with him. We set up a company by the name of W.E. Graham and Company, and then took over whatever accounts they had in the area and started expanding. The products were cleaning supplies, including machines etc., disinfectants, paints and varnishes, paper products, (towels, cloths.) We had our head supply depot and office in Fort William Ont. As we grew, we also took on some other lines; one that I recall was Jeyes Fluid from Jeyes of England, an extremely good product, had many uses.

When I started selling for W.E. Graham and Co., it was all on commission, we started off with 20% of calls except paper products which was lower. The first month my earnings were $175.00, I had to pay my traveling expenses including the car. Merelie thought we would be going hungry at those wages, however within four months I was earning $400.00 and by the end of the year between $1200.00 and $1500.00 per month, which was very good in the fifties for an average salesman. We also started expanding by hiring additional salesmen, which we took a 2 ½ % override on their sales and that improved our earnings. Wilf Graham was an exceptionally good fellow to work with, a real joker as well as smart and exceptionally good salesman; he and I had a competition going between us continually. We did have an advantage. We had the best products on the market and didn't have too much competition if we put the effort in selling. Occasionally we would go out to some of the outlying towns around the Lakehead together for two or three days and really call on every prospective customer, he would go one way and me another. We also had meetings in the evenings with school boards or hospital staff, we also taught people how to use some of our products and save them money. Once we had this system in any school or hospital, there was no competition as they didn't have the products or knowledge that we had available. We also used an 8 millimeter projector to show people how and why they should use our products and the results. It was a real joy selling this way and could talk to any prospective customer with confidence.

While we were living in this house on Hull Ave., our black and white Cocker Spaniel Clipper caught a disease, distemper and had to be destroyed, the whole family missed him as he was such a good dog. We did have a small red Cocker Spaniel named Kernel; he had been left with Merelie when he was about a year old by a fellow that hauled coal. Kernel had never been trained to live in the house and Merelie had quite a time training him but finally did with persistence. The final house training happened one day when we were at mother's, we had been there a few days, mother and Merelie went downtown and when they came back the dog had made a mess on the floor. Merelie was so angry; she picked up Kernel and literally threw him out the door. I guess that is what it needed to get it through his little head not to do this in any building, she left him outside all day and finally brought him in that evening, and from then on he was a perfect dog. The poor dog had never had a bath; we washed coal dust out of him for a week before he really looked like he should, turned out to be a real smart dog. Keith had a paper route and sometimes Faye would go along and help Keith deliver papers, Kernel would ride in Faye's basket on her bicycle and looked forward to this. Sometime Faye and Keith would go off playing in the neighborhood and leave Kernel at home. Merelie had trained him to call the children if she wanted them home. He would sit on the back step and howl real loud until they either answered or came home. This worked real well for Kernel because if he wanted to go and play where they were he would howl until they took him along. He was such a good specimen of a Cocker that we decided to get a female Cocker and maybe raise some pups. We bought a black Cocker female and called her Mandy, she developed into a beautiful dog. We did have two litters before having Mandy spayed. When we did get Mandy we had a doghouse outside and this is where she had her pups. I remember early one spring she had one litter and it was still freezing at night. One morning we came out and two of the pups were out of the kennel, they barely had their eyes open, however we put them back with their mother and they didn't seem any the worse for their experience. Mandy had five or six pups with each litter and each time there were two black and tan pups, real colorful, black body with tan legs. These pups sold quite readily, however we didn't want to have Mandy just for breeding, she was a good family dog.

In 1953 Merelie came into a few dollars, I believe it was $900.00 for an estate settlement, so she wanted to buy a car; there was an ad in the local paper for a Ford sedan one year old for $1200.00 so she went and saw the owner and looked the car over, it was quite dirty inside as the fellow had a truck and worked in the bush and used the car for hauling oil and tools and whatever. She offered him $900.00, he said no, he wanted $1200.00 as he wanted to pay the finance company off to clear his truck, so she left him her phone number and said this cash, in fact she showed him the nine $100.00 bills and said "this is all I have so think it over." A day or two later he accepted her offer, she brought the car home and cleaned it up, and it was really a nice vehicle. This was getting along towards fall, one day my sister phoned and had just come back from southern Ontario, and was telling Merelie about visiting our cousins there and about all the fruit that was going to waste as there was so much that our cousins left it to rot on the trees or ground. Merelie got all excited and said "I have my own car, why don't we take a week or so and go down there for a visit and bring some fruit back?" I was busy that day working on our roof, I think I had just finished putting a new roof on the house on Hull Ave. Merelie said we could leave this afternoon, so we discussed it and how about taking mother along as she was alone, as dad was away cooking. When we called mother she said she would go but wanted dad to go also, so we had to delay a day to contact dad who was about 200 miles east of Port Arthur and we would be going through that area, so we sent him a telegram and said to meet us at the depot in Longlac. When we met him the next day he was a little shook up as we hadn't told him in the telegram what we were up to, only that he should be prepared to be away for a couple of weeks. Dad could leave quite easily at that time as he was teaching camp cooks the art of cooking economically and cleanliness etc. for a large timber company. We met dad at Longlac Ont. and told him he needed a holiday, and away we went to Sarnia and Niagara Falls and points east. When Merelie said "how about going down there?" I said "we have never seen the falls so why not" and we were on our way. That was always the way we went some place, right on the spur of the moment and always enjoyed ourselves. This was a good trip, we did get some fruit and came back through Michigan, this was the first week in October, the scenery was beautiful, you have to see the color of the leaves in the fall just to see how outstanding the color is. There was a hunting season in that area for deer when we drove through the state. One small town we drove through there was a pub and we thought every hunter in the area was in that pub, you could hear the noise right out on the street and when we were out of the town about a mile there was a beautiful buck with a magnificent set of antlers standing by the road; it seemed as if he was thinking "where are all those hunters?"

The only problem we had with Merelie's car was the next day and it was Sunday, the universal joint on the drive shaft broke or was just worn out. We were short of money by now and wanted to get home as quickly as possible now, so we tried to find a repair shop that could replace the part and we could continue. After about half a day we finally did get a mechanic that found repairs for us and he did the work, it was evening, that night we drove quite a distance to make up the time lost. We arrived back home safe and sound after about 3000 miles.

We had a good trip with Merelie's car, she said "I'm going to trade this for a down payment on house for an investment," and what do you know, very shortly thereafter she made a deal on a real nice bungalow, only a few years old that the seller had built, he wanted to rent it back until he had another house ready to move into, so we agreed to that, however he had some teenage children and when they moved out, we had some repairs to do that wasn't there when Merelie bought it. After the repairs, we rented it out and it was a good investment; I believe a couple of years later Merelie sold it and made $5000.00 profit.

We were really growing with W.E. Graham and Company. Wilf was talking about bringing the product in from Huntington in Pennsylvania in concentrate and do our own mixing in Fort William as it would cut down on freight costs and give us a greater profit, and this we did, we had to start expanding across Canada to the west coast as our franchise was from the head of the lakes to the west coast in Canada. Wilf went to Winnipeg and opened a branch and asked me if I wanted to go to Vancouver and set up an agency and then start filling in back to Calgary etc. I thought it was a good idea, so trained another two salesmen to take my place. With these men working, I would have a good salary to carry on, with override on commissions and this was needed for going into a new area.

In 1956 we prepared to move to Vancouver, sold our home to an accountant for the Department of Indian Affairs, and had a nice wife and daughter 1 or 2 years old. Mother and dad continued to rent the basement suite. As he needed a mortgage, we carried it and made the payments the amount of the suite rent, which we thought was quite fair.

We bought 3 Plymouth station wagons, one for Wilf Graham, one for a salesman, and one for ourselves. This was the first new automobile we had ever owned. We had a trailer and heavy hitch put on the rear of the wagon to tow a trailer I had previously built which was a good sized 2 wheel trailer and loaded with everything we felt we couldn't do without, also sold our furniture and gave Kernel, our red Cocker Spaniel to Wilf Graham, as he liked him and had a good large yard for him to run in, Mandy, our black Cocker, we left with my brother and his family until we would get settled and then we would send for her.

There were quite a few parties held for us here and there before leaving and were presented with a real nice briefcase with my initials on from Wilf and Clarice Graham. We left the third week in June 1956 as soon as Faye and Keith were able to leave school. Something else we did was sell our hunting rifles and bought a good camera and planned on hunting with the camera in the future. We traveled via the U.S. highways 1 or 2, then on to Yellowstone Park, we would stop whenever we found a good place to park and we all slept in the station wagon, in those days it was quite safe to do that, today, I wouldn't even think about that with all the kooks that are on the highways. Well anyway, one evening it got dark before we found anywhere to park, we were out in the prairies somewhere, anyway everyone was getting tired, we came to a turn in the highway and could see a nice grassy wide spot ahead by the headlights so here was a place to park. So everyone got their pajamas on and climbed into bed, had a real good sleep until about 4am, and all of a sudden it seemed as though every heavy truck in the U.S. was driving past us. When we woke up it was quite light and what do you know, we are parked in a diamond with three highways on all sides of us, and all the transports were really rolling. There was also a railway track running nearby. We never forgot that stop and always made sure we stopped in the daylight from then on. When we went through the Yellowstone Park we spent some time looking at the sights and taking pictures with our new camera. Faye and Keith each had a camera also. One place we stopped at was Old Faithful, the water spout that shoots in the air every hour or so. Faye and Keith had decided or it might have been Merelie or my suggestion that they don't duplicate their picture taking and then they could compare pictures at the end of the trip and have extra pictures made and this would save them money for their film. Well anyway, they couldn't figure out who should take the next picture and old Faithful made its spout and quit before either of them took a picture. Needless to say we had to wait around for Old Faithful to spout again, and this time I think they both got pictures. When traveling through the park there are signs "don't feed the bears" and "keep your windows closed in your automobile," up ahead of us, several cars had stopped while someone not paying attention to the signs was feeding a bear and there was another bear, about a 2 year old black going alongside each car and checking to see if there was a window open or any handouts. Merelie and I saw this bear at the car in front of us and said to Keith "close your window" as he had it open, but before he got around to winding his window up, the bear had come along beside us and stood up, poked his head in the car and a long tongue came out of his mouth, went slurp and almost licked Keith's face. Keith was fast then rolling up his window and almost caught the bear's head in the window. After that, when we said to shut the window, there was never any delay.

We arrived in Vancouver about the end of June and drove right through town to the waterfront as we wanted to catch a ferry to Victoria, it was getting dark and of course we didn't know where we were, so we drove into the first garage we saw and asked if we could park our trailer for the weekend. There was quite a chuckle from the fellow we asked, he said "well you certainly brought your trailer to the right place, this is the city police garage." He spoke to one of the other fellows and said "sure, just back it over here, but be sure to come and get it next week." So off we went with his directions to catch the C.P.R. ferry June 30th 1956. We boarded the last ferry about 9pm and stayed on board all night, sailing from the waterfront at Vancouver to the dock at Victoria harbor early in the morning. We met my brother Lorne and wife Phyl. We went in their vehicle for a tour around the downtown section, then out to their home for a visit for the weekend. On Sunday we went up the island to Duncan and had a picnic on the Chemainus River. While there Faye went in the river swimming and we coaxed Keith to go in also, we bet him 25c that he wouldn't go in, he was a little timid about swimming, however he went in and got caught in the current and almost drowned if it hadn't been for someone catching him down river about a hundred yards or so. After we were settled in Vancouver, he was enrolled in the Y.M.C.A. and learned how to swim properly and has enjoyed it ever since and became a good swimmer.

We left Faye and Keith with Phyl and Lorne for a couple of weeks and went back to Vancouver Tuesday morning to locate a place to live and a warehouse for supplies to start our sales promotion on the west coast. That was on July 3rd /56. We rented an apartment on the corner of Vine and Broadway in Vancouver and warehouse space in an old building on 4th Ave. where there were railway tracks, but for 2 or 3 weeks we had a small place with Mrs. Brice until we found the apartment, first we were on Kitsilano beach. We were driving around Vancouver and eventually picked the Kitsilano area to live in also for our storage area for the company. We really did cover all of Vancouver and surrounding area to get acquainted with the area. Finally we saw a sign on a house near Kitsilano beach went in and met the lady who owned the house; her accommodation wasn't suitable for a family but would do for Merelie and me until we found something for the family. Mrs. Bryce was really happy to have us in her home until we could get something larger. She was a very hospitable and happy lady who needed some extra income and we became fast friends. When we found our own place we'd visit back and forth and later she came to visit to the resort we owned in later years. She was quite a joker and tease and was always saying some day she had to find another man after her husband passed away which was shortly before we met her. One time when she came to our resort, she had a cabin for two with a double bed, so we arranged for a young man to hide under the bed until she was settled then come out, as she was looking for a man, so he had heard. She thought it was quite a joke and we all had a good laugh.

One time during this period, someone had arranged for her to attend a dinner at a Mennonite farm in Alberta. While she was visiting some of her family, they had also made it known what a joker she was so had some of the men at the dinner get up and go around and look her over as if she was a piece of merchandise for sale, they didn't talk to her but made her wonder what was taking place, so one of her friends who had arranged this charade said "didn't you know this is one of the reasons we were invited to this dinner, you have been saying some of these days you would have to get another man and these men looking you over are also looking for good healthy work mates."

I started calling on prospective customers, and found that in the Vancouver area, there were a lot of buyers who didn't want to change or were real loyal to the company they were now buying from. While in Victoria about three weeks later, I did get a good sale and steady customer with B.C. Electric. Their manager was willing to listen and was really happy with my service, however it was a rough row to hoe and sales came in very slowly; later I found that Huntington, our major supplier had a representative in North Vancouver who had most of the hospitals and good accounts and I'm trying to sell the same products with a different name, as our company distribution name was W.E.Graham and Company, a very difficult situation. Huntington was supposed to have phased out their outlets, and we would take over but that wasn't happening. I also tried to hire salesmen to start on a straight commission and one fellow started part time but soon gave up, however I found that in the Vancouver, the one looking for sales positions wanted a salary and small commission or they wouldn't start to work, where as the area I came from you really got out and worked and it paid off.

We really liked B.C. and enjoyed living here, there was so much to do, and we made some real good friends. Merelie joined the Rebecca Lodge and I joined the Odd Fellows Lodge, through these we met more people and became active in the Lodge and community. Faye and Keith were well settled in their school, we also took part in the local church, and eventually Faye and Keith attended Sunday school activities with the Salvation Army who were very active in our area.

We did our best to build up a business for a year and four months, and by this time I had one man working part time, so decided to go back to the Lakehead for six months or so to train some salesmen there, leaving our part time man to look after our accounts in Vancouver.

The Plymouth station wagon was one of the longest vehicles that Chrysler Motors ever manufactured, I could never find a place to park in the downtown Vancouver area, one day I drove around for 4 miles trying to find a place to park; that was it, I came home and started looking for a short car and eventually bought a Volkswagen beetle, that was short enough to park, then sold the Plymouth wagon.

We left Vancouver and moved into a house in the Fort William area. There were not many places to rent so we had to take what we could get and pay a high rent.

I must tell you about an experience we had while driving east through Saskatchewan in our V.W. bug. While driving through the prairies we would drive as far as possible before stopping for the night and in this case it was now dark and we were out on the highway on a lonely stretch which we shouldn't have been at that hour with the children and whatever we could carry. A vehicle with one man in it pulled in front of us and tried to flag us down. Merelie was driving as we would take turns, to be able to rest. We didn't like the looks of the situation so we didn't stop, this fellow tried to stop us several times, and we finally pulled off the highway into a store and went and got something, figured we lost the vehicle who tried to stop us, however he was parked in the underpass which we had come through right in the middle of the road, and tried to stop us again but we squeezed past him and sped on for the next town, he tried to stop us again once or twice more, however we pulled into a gas station that was open, there was an attendant and an R.C.M.P. officer there. We told them what had happened and the policeman said "I hope you have the license number." In our excitement we had not taken the license number, we knew the color and make of vehicle, what the fellow looked like, but no number. The officer said there had been a number of robberies of gas stations and holdups in that area for the past 2 to 3 weeks and suggested we stay in town until breakfast, which we did, we picked the first place we could find, which wasn't the best. Faye and Keith and I went to bed but Merelie couldn't sleep and sat up in an easy chair as she was too nervous. We shall never forget the name of that town, Moosamin Saskatchewan. We were carrying quite a lot of money with us which was a good lesson and we haven't done that again under those circumstances. We arrived in Fort William and Port Arthur the next day, after a visit with friends and relatives we rented a house in Fort William and arranged for Faye and Keith to go back to school. The spring of 1958 Merelie and I staked mining claims of uranium but couldn't mine these as they were in a park near Port Arthur.

Due to the slowness of being able to get clientele and business rolling in Vancouver, I continued on at the Lakehead to train salesmen and maybe take some help back to Vancouver.

I bought a 30 30 Winchester and went hunting with Wray. We got a moose for our winter's meat, also had been out with Keith, dad, and

Lorne and got some partridges. Lorne broke his stock on his rifle trying to hit a partridge.

Our intentions were to stay through the winter at Fort William then go back in the spring to Vancouver. We stayed a year and a half then went back in the spring or early summer of 1959 as soon as school closed. If I remember correctly, the winter of 57-58 was one of the coldest on record and we began to think why stay in this cold area, with the milder temperatures at the coast. With determination to make that business profitable; it wasn't that purchasing agents didn't accept us but B.C. people are very slow at changing regardless of how much one talked, many prospective buyers looked forward to my call however they said "we want to make sure you are going to be permanently established before buying your products. How was I going to be permanently established if I couldn't sell them some of our products? Anyway I hung in there, was getting an override commission from Fort William sales which gave me the support I needed. I know now I would have done much better if I had made more calls through the interior as there were not many salesmen traveling through to interior areas at the time.

When we retuned to Vancouver we bought a house on Trafalgar St. in the Kitsilano area and made up our minds we were staying, rain or shine, which we have however with quite a few changes. I stayed at selling and business was not booming by any means, however the B.C. economy was growing, and by the late winter or early spring of 1960, had made up our minds to venture out in something else, had one part time salesman to look after what I had established and started looking for something to invest in. talked to a friend we had made in the real estate business and he suggested looking for older hotels or various types of business that could be bought, spruced up and resold at a profit, this sounded good, so off Merelie and I went. It may have been the spring break when we could take Faye and Keith with us. Anyway we made a tour up through the interior on #3 highway; saw a hotel at Grand Forks, as it was listed with our friend in Vancouver. We came back in a circular route through Kamloops and down through Merritt and Princeton. We also found a resort and ranch for sale on highway #5A on Allison Lake about 20 miles north of Princeton, which really impressed us; as it turned out the hotel had been sold while we were traveling, so we discussed the purchase of the resort and ranch. We were going to have 3 partners in the project and therefore would not be short of capital; I would give up the W.E. Graham sales and we would move into the resort and operate it, our partners were in agreement, so we made an offer. We made the mistake of just putting our own money down instead of so much from each partner, our offer was accepted, now we came back with our hand out for a quarter of the price from each supposedly partner. The first one, our realty friend had just bought an apartment and was short of money for a few weeks until it sold, the second, another real state salesman had no money all of a sudden as his wife had left him, took all of the money with her back to England, number three, his wife was ill and didn't want to move from Victoria at that time as he was also going to be a working partner with a quarter of the money in. Now here was Merelie and I with a $10,000.00 deposit and not enough to cover our obligations, or lose our deposit by default. Well anyway, we traded the house as part payment, obtain what we could from insurance and had a mortgage at $10,000.00 a year @ 5% with 2 payments at $5,000.00 July 1st and October 30th. We took over that resort with less than $2.00 to put in the till of the store.

We never looked back, although we did have to borrow some money from a friend for our first payment in July for a few days. We took this resort and ranch over in April 1960 and operated it until December 1st 1969. We had a lot of experiences, trials and tribulations during those years but we survived and were financially better off, however our health had somewhat deteriorated by 1969 and we needed rest from too many long hours and not enough rest.

When we bought Skye Blue Lodge resort and Ranch, Merelie and I looked like a couple of kids although our son Keith was 17 and daughter Faye was 15, there was lots of work. Merelie took over the resort and I the ranch and Faye and Keith as helpers plus we had 2 or 3 employees. We had 19 cabins including a log cabin which we used for help accommodation. We had a store and restaurant in the main lodge which had a large recreation room and the largest fireplace you ever saw and 3 or 4 rooms upstairs, a garage and workshop, also a pump house with a pelton wheel that generated our electricity. We also had a Caterpillar generator for backup power. The pelton wheel and water supply was quite unique, the water came from a small creek with a catch dam about a quarter mile up in the mountains. The ranch consisted of 450 acres of pasture, hay fields and side hills with pine and fir trees, about 170 acres of hay fields, 20 acres of pasture, hay barns, one the old original, and the second a new barn that held over 1000 bales of hay in the top storey, some farm machinery and 12 head of horses for trail rides, as well as 26 rowboats and a Jeep for taking guests fishing to some of the mountain lakes, also for ranch work.

We also acquired an old dog named Knight who had been there most of his life and the previous owner didn't want to take him away from his home; he was a good old dog. We got quite a charge out of him as he had learned that if he brought some wood for the cook stove he would get something to eat. We had Knight for about 2 years when he finally became paralyzed from age and overweight and we had to put him to sleep. As to the horses, there were 2-3 year olds that had never been trained for riding, 6 were middle aged and 2 were old or very old. Shortly after we bought the resort we had the oldest taken away as it looked awful and would soon die anyway, then shortly after we had to dispose of one of the riding horses because it was stifled and could not be ridden for more than a few minutes or it would go lame, a nice looking horse which we had all taken a liking to and were sorry to see him go. Now we had six horses suitable to ride.

This whole operation was appropriately named Skye Blue Lodge and Ranch as all the buildings had blue roofs and oil stained walls, was certainly a beautiful setting on the end of Allison Lake with blue water. Skye came from the Isle of Skye in Scotland where the previous owner had bought Hereford cattle.

We could accommodate about 85 guests plus we had lots of room for parking RV's, which started coming in shortly after buying the resort. Two or three years after buying, one early morning before guests started moving around, we took pictures of the resort with waterfront lined up with RV's in 2 or 3 rows estimated between 35 to 40 units. This really boosted the restaurant and store also horse rides, boat rentals, also kept the Jeep busy taking fishermen to some of the upper lakes. There are over 30 lakes in that area which was kept stocked with rainbow trout, however that is another story which we will go into later.

Shortly after moving into Skye Blue, an old Swede fellow came along one day and asked if we needed any help as he was looking for work, and as it turned out, we had been told about him from the previous owner's son-in-law, and he said if Charlie comes looking for work take him on, it will be the best investment you will ever make. Well we hired him. Charlie Lofquist, he was a real rugged man in his early seventies and in better shape than anyone at fifty. Charlie worked for us mostly summers all the time we had the resort. One year, I believe It was the second year, he looked after the resort and fed the stock while it was closed after the Christmas holidays while Merelie and I went for a holiday down to Arizona and California, however we never wanted to do that again as he would usually go south to Yuma Arizona for a holiday in the winter and besides, he said it was too lonely for him when no one was around. Hurrying didn't always pay either. When Keith had finished with his own work he would help out in the store, pump gas, take trail rides out, and at times when we were without a cook, he would help in the kitchen. The whole family did all of the jobs at one time or another to take up the slack or if another person wanted extra help. We all became versatile at any of the work at the resort and ranch and soon learned how to do everything.

When we took over the ranch, it was producing about 125 tons of alfalfa per season with 2 cuttings. One day while looking over the fields after our first cutting, a guest from the resort walked out on the fields where I was and asked what I thought of the crop. I said "it should be much better but wasn't sure what to do." As it turned out, this man was Dean Clements, a retired U.B.C. professor of agriculture who was well acquainted with this ranch as he had been a friend of Mr. Anderson who had been a former owner before his death, after which Mrs. Anderson had let the ranch go and had even lost their cattle grazing rights. Dean told me what to do to bring the crop back to full production, we need soil tests to find what the ground was lacking in nourishment to produce the alfalfa; he instructed me to take soil samples all over the fields, then divide it into three bags and send one bag to the Dept. of Agriculture, one to the dean, and one to where we may purchase fertilizer, in this case we chose Buckerfields who was one of the larger suppliers. When the soil test results were returned, the Dept. of Agriculture and the dean were the same, while Buckerfields was different. When I questioned the dean about this, he said no doubt one company had an over supply of a certain fertilizer and were trying to pedal it off on an unsuspecting prospective customer, needless to say, that one company never sold us an ounce of fertilizer. Because we needed so much to start with, I talked to a local feed and fertilizer supplier in Princeton. When I told him I needed over a rail car load, the local supplier couldn't believe it, in fact he tried to talk me out of ordering that amount, however when he finally realized I was going to buy it anyway, he ordered it in. There is one thing I wanted from him was a discount on the price for this quantity and when the crop really started to improve and other ranchers started to buy fertilizer, I would get a cutback or a payback for showing what really could happen by fertilizing these hay fields . The local dealer tried to tell me the fields would all be white and kill all the alfalfa, however the dean was sure he knew what he was doing as in previous years while Matt Anderson had been alive he had experimented with these fields and found that they would really grow good crops with that type of soil, fertilizer with the right proportion and mixture with plenty of water which we had through irrigation.

  When our second crop came up after the fertilizer, it had improved considerably but the following spring we put on the same amount of fertilizer before the first crop. On the second year we doubled our tonnage, now we were on our way, however we still had a problem with pocket gophers in the fields. These little varmints dig holes in the fields and push up large mounds of dirt. In our case, with stoney ground, these mounds were any height from 6 to 15 inches. We were cutting our hay at that time with a mowing machine which consisted of a 7 foot bar and sickle which ran about 4 inches from the ground, the stones would jam in the sickle guards and break the teeth of the sickle. The first cutting season it cost us a bundle just to replace sickle blades and also so much time on repairs. Eventually the Dept. of Agriculture came up with a machine to poison these varmints.

Faye and Keith were going to school in Princeton but also helped around the resort and ranch. Faye helped in the store or took trail rides out. Incidentally, we had to buy several more horses to make up for the cripples we had to dispose of. Faye was real good at this and took over real well especially during the summer months. Keith liked the ranch and was a real helper with the irrigation pipes that had to be changed or moved twice daily and this was no little task; it took about an hour and a half morning and evening for two persons and lots of times Keith did the irrigation on his own. He was also good with the haying equipment, he took tins of oil and looked after things which I didn't appreciate as I often thought he was going too slow, but later when Keith wasn't there I found out that the ranchers all through the interior of the ranching areas of B.C. were all having their problems about that time, so the department rented the machine which deposited several kernels of poison oats about 8 inches under ground. This machine cut a trench about 12 to 15 feet apart. It certainly eliminated the problem in one season. The ranchers had really caused the problem themselves by throwing the environment control system out of whack. For years the ranchers had been killing off the coyotes which had been controlling the gophers, but as it turned out when this was realized the coyotes were allowed to multiply again, there were not so many pocket gophers, in our case we would not allow anyone to shoot a coyote. While raking and baling hay there would be 2 or 3 coyotes running along with the rake or baler and would be catching mice and gophers continually. If the tractor stopped, they would stop and when you idled the tractor down, the coyotes would run off into the bush until the tractor started to move then they would follow again. We only had to poison the gophers one season to eliminate the problem.

As to the resort, we enjoyed meeting most of our guests and many became good friends and we looked forward to seeing them each year. When we first moved into Sky Blue we had some of our old friends come for a visit but eventually they stopped coming because we really didn't have time to visit during the busy summer season. We did finally get some to come in the spring or fall when everything wasn't such a beehive of activity.

We had to hire a cook for the restaurant for the summer season, sometime Faye or Keith or myself would take over the cooking, this was one thing Merelie wouldn't do, she had had cooking up to her ears when we had the bush camps and wasn't about to get hooked into this again and beside she did not have the time with all the activities that had to be looked after in the resort. There was the registering of guests coming and going, the laundry, the store, and all the activities that our guests wanted some part of. We hired help, had a woman for cleaning cabins, helping with laundry also one for the store and waiting on tables in the dining room, also a maintenance man to repair the rowboats also painting them as well as painting or oiling the cabins which had to be done every year.

The first spring we were at Skye Blue some of the local ranchers wanted Faye and Keith to join the 4H club, purchase 2 beef calves for their summer project. Well they each brought them home and they each bought so many pounds of grain so they could be fed the best and become prize calves. We put them in a corral and suggested they start to teach them to lead and get used to people handling them. Keith started off, he got a halter on his, however it led him or dragged him around the corral through the manure and he was an awful mess, but he stayed with it and finally taught his calf to lead. Now as Faye saw the problem Keith had, she decided she wasn't about to teach one of those calves anything, and she said to just turn it loose, she wasn't interested, so eventually Keith taught both calves to lead, however Faye's calf didn't get any grain or tender loving care, it was in the pasture with the horses and other cattle all summer. When fall came around, both calves were shown, and Keith's calf came in 3rd and Faye's, with no grain and not too much attention came in 9th I believe, however there were a lot of calves that looked as good as either of our calves; there was one thing that it proved, was that it didn't pay to feed grain if a calf had good pasture, I mean this cost wasn't profit wise at that time, These calves were bid on by large food stores at that time.

The first handyman

we hired was from Calgary, he said he needed light work as he had been injured the previous summer and was still recuperating. As it turned out we had been watching the chuck wagon races the year before at Calgary when a fellow was seriously injured, the horses and wagon went over him, he was taken away on a stretcher, here was the same man, he had scars around his back and sides where the doctor had opened his skin to repair his ribs and internal injuries. He was fortunate to be alive but was a willing worker and was with us for a good part of the summer.

When we bought Skye Blue it was strictly a summer operation, however a couple of years later we had a few guests for Christmas week, this was the start of our winter operation which grew each winter. In the spring, fall and winter we had time to spend with our guests and they became good acquaintances and friends. We still have some of these people for friends over 35 years later, long since we sold the resort. During the fall we would get hunters and fishermen or people just out for a walk and enjoying the fall colors, winter, it was for fun in the snow, ice fishing or skating on the lake also. We started to fix up a ski hill but never did get it completed before we sold, also planned on a bobsled run. As I said before we had a lot of fun and experiences while at the ranch and resort.

The second year we were there, we bought a 2 ½ year old black and white pinto stallion colt in the fall. The spring of 1963 we had our stallion who we had named Chief, registered name was "Little Chief Big Heart," bred a nice quarter horse mare, and in 1964 we had our first pinto colt, dark red and white, a beautiful animal. Back from the lakefront we had 20 acres of pasture where our horses and a few cattle grazed, and this is where our colt grew up. We named him Apache because he was the only colt that year, everyone made a fuss over him, and the way he behaved with people we often thought that he thought he wasn't a horse but wanted to be one of the people, and if we let him, he stayed right near or with anyone who would pay attention to him. We were certainly pleased with our first colt and of course had Chief breed all our mares so the following spring we had half a dozen colts, all pintos but one which was black with a white forehead and socks which we named Poncho, which was the proper name for him as he fit the name as he later developed. The rest were red and white, well marked, there was one out of a palomino ¼ horse mare that was tan and white, anyway we were well pleased with our stallion as he had strong genes and was throwing almost 100% pinto foals. Most of our mares were solid color, one mare named Flash was a black Morgan mare. We also had a palomino ¼ horse, one thoroughbred cross, a light buckskin, one dark brown ¼ horse, a bay ¼ horse, a ¼ horse buckskin, one ¼ horse and pinto. We usually had 4 or 5 foals each spring, all pintos except 2 blacks and one buckskin, very nice horses and colorful. We sold some of these pintos and one went to a local rancher who said it was one of his best for use with the cattle, incidentally, our first pinto Apache would spend his leisure time herding the cattle in the pasture he was a natural cow horse. When we started to break him for riding, we would put a horse blanket on him then walk him around for a while then put a light saddle on and walk him around again and when he got a little stronger we would put hay in two bags then put them on each side of the saddle and walk him around some more, and when he got stronger we put a light person on him and eventually an adult. He never once bucked anyone off. When taking him out on trail rides, if a person said that he was a good rider we would take him up to Apache and if his ears went down we knew right away that person couldn't ride. Keith rode him most of the time. His dam, Flicka was real good for cattle work and no doubt this is where Apache got it. Flicka always liked to have lots of attention when she would have her foal. Several times some of the guests saw her foaling which is different from most animals and the other mares usually went off somewhere on their own, and this almost always happens with beef cattle. When we bought Skye Blue each of the family claimed their own horse. Dad had Chief, mom had Flash, Faye had the horse that was lame and we had to sell, and Keith had Flicka. When Apache, the first foal was born Keith gave back Flicka and claimed Apache. We sold quite a few colts, the prices were really good. We used the horses for trail rides all summer then in the spring the mare would have a colt which sold for $175.00 up to $500.00, two colts went to Winnipeg for mixed running race track from our horse Penny.

At the same time we were only getting 17c a pound for young beef. We did have one advantage for some of our beef; we raised them in the pasture, had our own cooler and would butcher the meat for use in the dining room and café. We had many patrons come back for one of those steaks, and they would comment that it was the best they ever ate.

We were fortunate that we never had any serious accidents with our horses; of course they were well broke, very quiet and docile, even our stallion was very quiet and liked attention. We would sometimes tether Chief out near some of the cabins to eat the grass especially around the Kiddies Korral, the children's play area, and it wasn't long before Chief found that these children and their parents had handouts for him if they could pat him. We have seen him spot some people and he would walk back away from them to get his chain behind him then walk forward, this way, no one got caught in his tether chain. This horse was really good to ride; I've spent many pleasant hours riding him. We used him for lead trail horse quite a lot, also for cattle and horse round ups, he was very sure footed, and I have ridden him on a lot of side hills while at the ranch.

Occasionally we would hire someone to train our horses of mostly over 2 years old to 3 for guests to ride. One spring a young fellow came along from the Fraser Valley, he said he wanted some work with horses, but I was doubtful of him as he appeared to be a smart aleck type of fellow, he said he had worked a couple of summers for one of the ranchers in our area. This rancher was hard on horses, so this didn't impress me either, anyway, I said we would try him for a day or so and see how he made out, told him what we wanted, quite non spookable saddle horses for guests. The first morning I let him work with two 3 year old pintos in the corral, this way we could see how he handled the animals, he seemed to be quite gentle with them. By noon he was riding one around the corral quite well, however the second one bucked him off a couple times. This colt was out of a good size ¼ horse named Cindy, she had a mind of her own and it appeared her colt was the same, he was a nice looking gelding pinto and a good size, so maybe he could make a cow horse out of him, which we did and sold him to one of the ranchers in our area and once we had him properly broke or trained, which we did with all the foals we kept to 3 years old. The one we liked riding was our first colt Apache, and after riding him several days, he came back to the lodge from one of our lake trails with an old mountain sheep on a lariat tied to the saddle horn, a wild one with a nice set of horns, we said "what are you doing? We can't have a wild sheep in where the guests are." We were really surprised to see this, especially with a young horse. He said "when I was in at lunch time I told you all that I saw some goats or sheep or something out on the trail and you all laughed at me, well, there is one of the animals I saw and what is it"? We told him it was an old ram mountain sheep which we hadn't known was in our area, it was very old no doubt or he would never have been able to catch him. We told him to take it back where he got it and turn it loose; it also had a very strong odor. We later talked to the local game warden, and he said there was a small herd in the area, and when we told him about the incident, he couldn't believe it. This same young fellow was riding another one he wasn't training when we saw him standing on the saddle riding through the yard, there was one thing for certain, any that he trained were not spooky, as he had done everything with them.

As you well know, we had a lot of funny things happen. One spring Charlie wanted us to buy a milk cow, well I wasn't too impressed because I wasn't going to milk a cow any more, so Charlie said "you buy one and I will look after her." We did buy a nice small Holstein and Jersey cross which just had a calf. We had lots of milk and Charlie took care of the cow and milking, and he made a real pet of her. Charlie milked her at 6am and 6pm and it wasn't long before Daisy, as Charlie named her, that she was at his cabin door bawling for him at a quarter to six every morning if he happened to be a little late, and if at some times she was late coming in from the pasture, Charlie would go out and shake the milk pail and call Daisy in his Swedish accent which sounded more like "Daysee" and she would come on the run jumping and running like she was crazy. One day I heard an odd noise out behind the barn and went out to see what was going on, well there was Charlie holding onto Daisy's horns pushing her backward and shaking her head, then she would push him backward shaking her head, and both of them were making grunting and snorting noises, well I watched for a minute or two then Charlie saw me so I asked what was happening. Charlie said "Daisee has no one to play with so I am playing with her den she won't be lonely." So that was that and Charlie and Daisy had a push and shove match every day or so as long as we had her which was several years.

There were many things that happened over the years we forgot which year something took place. One spring when we had some foals, I had the horses in the hay field beyond the regular pasture area, and as I heard a commotion with the horses, I walked out to see what was going on, this was about an eighth of a mile from the barns, what was really going on was an old sorrel mare named Queenie was trying to take a foal away from a young ¼ horse mare named Teaka, and there was a fight going on between the two mares, so I tried to stop this and finally opened the gate to let Teaka and her foal out, well, all the horses rushed through the gate and started to run towards the barn, I had left the first gate open when coming to see what was wrong and now had to run to close that gate or I would have horses running everywhere, well I'm running for all I'm worth, the horses are running and jumping and squealing and what would you know the milk cow Daisy gets in the act also, she is running and mooing behind me, and all of a sudden I thought, if someone saw this menagerie with me in the lead, they would think I had gone crazy. I know it must have looked unbelievable at least hilarious.

We had quite a few mule deer in our area; it was quite common to see 10 or 15 in the fields and quite often several right up to the lodge. One day some American tourists came into the café for lunch, and some of them said, as I was walking through the café "do you ever see any deer or any wild animals?" I had just walked out of the lounge area and saw some deer feeding on the lawn, so I said" oh yes quite often, go out in the lounge an you will likely see some deer," so out they went and couldn't believe what they saw, there were five deer right in front of the lodge feeding and not paying attention to anyone. One year a doe had twin fawns, she was quite tame and would come down off the side hills to feed and we would see the two fawns partially hidden behind a tree with a head sticking out on the other side watching what was going on around the resort.

Merelie was always full of mischief, so one time some American tourists came into the café and asked her if there was much wild game in the area, and she of course said "yes there was," and they said "we didn't think there were lions and tigers," but they had seen some between Merritt and Skye Blue, so Merelie went along with their thinking and said "oh yes you never know what you might see." These people went away without a doubt in their mind that this was a lot like Africa, however we later heard that day that these people had no doubt seen tigers and lions, as a circus truck had gone over a bank on the hill south of Merritt and there were circus animals running around loose.

We had quite a few retired couples that came to Skye Blue, one particular couple were quite interesting, he had a hearing aid and liked to fish, so they would rent a row boat and go out in the lake fishing. She did the rowing and they were usually there in the fall when there were less people around, and it would be much quieter. We would hear him hollering to his wife to row harder on the right then more on the left, so after while you would see them just drifting, and he would holler again, then she would say something to him and this would go on every day they were there. One day Merelie said to her "don't you like fishing? You could row, "no he wouldn't do that, if I say anything to him while fishing he turns his hearing aid off, then keeps on hollering, a little to the right, a little on the left."

We had a very nice lake at the resort, also had a few smaller lakes in a fifteen mile radius which the government in those days kept well stocked with rainbow trout, or we also used a common name for that area, Kamloops trout. We have caught or had guests catch these trout over 22 pounds. One couple of fellows that used to come every summer from Seattle and who worked for Boeing there would always have a competition going on as to who would catch the largest fish. One year the pair caught 3 trout 18 to 22 pounds plus, well, they had a little celebration and the other fellow went out the next day to beat his partner, and I think he must have still been celebrating, as he upset his boat, lost all the tackle and almost drowned, but they went away happy and were back the following year.

Every spring we would hire summer help, first the cook, then waitresses for the café, which was quite busy in the summer, then a handyman to do repairs and painting. We repainted the blue roofs on the lodge and cabins as they had to be repainted every other year, the walls were all stained with boiled linseed oil, as the raw linseed oil would turn dark, our outside walls looked as though they were varnished, then we always had Charlie as a general all around helper. He had worked for the previous owner for several years and knew more about the place than we did. During the haying season, we had to have one man for changing the irrigation, which our son Keith looked after for the first three years, he then went to work for one of the mines in the Merritt area, and we had to have someone else, and that was my fault, which I will go into later. We also needed a person to look after the horses and trail rides, usually a girl as they were better wranglers than young men, and we usually needed the men in the field. We did have an older middle aged fellow one season that was really good. Our daughter looked after trail rides and horses for two years, then she got married and her husband was afraid of horses. We also had to have someone in the store and to meet our clients and get them checked in. We also hired a cabin girl to clean cabins and look after the linens and laundry. When we started haying in June, we always hired two and sometimes three men to handle the hay and store it in the barn, and in later years we tried to sell the hay out of the field, which cut back on men. By the 3rd year we were putting up, for sale, 600 tons which went to the Fraser Valley farmers and race horse farmers, and to some of the local ranchers. We bailed the 1st and 2nd crops , and our 3rd crop went into silage which we fed to the cattle and horses during the winter months. We gradually built our cattle up to about 160 head and we usually had up to 30 head of horses, and one year we fed 125 yearlings for one of the ranchers and we always had our milk cow daisy.

In the late fall and winter we would look after the guests ourselves as the winter business was spotty, a rush at Christmas and New Years then slower the rest of the winter. We always hired a couple of men to help us put up ice, which was used for the fishermen and the ice boxes in the cabins. We stored the ice in one end of the old barn and covered it with sawdust, the ice would keep all year. We had all ice box fridges when we bought the resort, gradually changing over to electric about our 3rd and 4th year, but always needed some ice. The lake would usually freeze over just before Christmas and everyone would enjoy skating. One year there was no ice on the lake until the night of the 24th of December and on the 25th, the skaters were chasing fish that they could see through the ice. Most years the ice would be 24 to 30 inches thick by February when we would cut and store our ice. We would cut it into 24 inch square blocks with a chainsaw, but prior to the chainsaw, we used a 6 foot saw which was made for cutting ice, one man would pull and push on the saw all day long; then the chainsaw made it much faster and easier. We would haul the ice off the lake with a team of horses and a sleigh to the log barn. We used ice tongs to unload and move the ice around. I mentioned a team of horses, most of our horses were trained with harness as well as saddle and could be used for either type of work, and during the winter season we always had hay rides on a sleigh with a team of horses which the guests really enjoyed. One winter we did have a young fellow working for us during the busy time. We taught or thought we had taught him how to drive the horses for sleigh rides; however it later turned out that he wasn't prepared for horses that were feeling good and hadn't been working for a while. One evening he took a sleigh ride out and everyone was having a whale of a time and I think the horses got in the act also, as when they backed into the driveway, the doubletree came off the sleigh which would happen if the tugs were not fastened up tight enough. When the horses broke loose of the sleigh, they took off for the barn with our young fellow dragging behind still holding the lines, as we had told him never to let go of the lines when you are driving these horses. Well he hung on alright although he didn't have any control of where they were going, and the end result, the horses jumped over the hood of a guests car and of course did some damage to the car, which we were not happy about, however the owner thought it was a big joke that evening, however the next morning he came to enquire about what had happened to his car. As it turned out and unknown to us, he had been drinking prior to the hay ride and didn't remember anything about the horses, of course we had insurance and had his car repaired or at least paid for it. You had to be prepared for anything when using horses, and after that we always drove the horses ourselves or made certain that someone knew how to handle and drive the horses.

When we bought the ranch and resort, there were two 3 year old quarter horses that had never been broke or trained for saddle, a mare named Teaka and a stallion named Tango. It had been a while since I had ridden any horses and there was no one else to do it, and at that time we couldn't afford to hire someone, so I started working with the two of them. The proper way is to get them used to being handled and used to equipment such as the bridle and saddle, also we would sack them out, which means using a cloth bag and swinging it around and over the horses body and really getting them to know that they won't be injured, then would start driving them around with a pair of lines and by doing this they became used to the harness which we also used, especially the horses we may use on a wagon or sleigh. As it turned out, the mare Teaka was a little spooky when I started riding her; actually she was nervous but never did throw anyone off. Tango on the other hand didn't mind the saddle and was not the least bit nervous but was used to throwing someone or as well as bucking, he threw me off, then the second time he tried, I swung him around so he couldn't buck, so then he started to run and would have run into the fence but I jumped off with the lariat in my hand, and when I jumped off he turned and started running towards the pasture, I threw the lariat like an old pro, really surprised myself, and really surprised Tango as when the lariat dropped over his head, I ran to a post and put a couple of turns around the post and hung onto the rope. When he came to the end of the lariat, he almost upset himself; now I knew I had to have some kind of a bit in his mouth that I could stop him. So I got a bit that was a little more severe and put it on him, and he seemed so quiet and docile when you were around him that I wondered why he wanted to buck, so the third time around I put him in the corral so he couldn't try to run away when I got on him. After getting in the saddle, he just stood in one place, looked to one side and then the other, and when I nudged him to move ahead, he started into his buck and I jerked up and sideways on the reins and he reared up and was going over backwards so I piled out of the saddle and tried to jump away from him as he was coming over backwards, my foot caught on a root and I fell and he landed partially on my legs but jumped up quickly luckily I wasn't seriously hurt, the Lord was looking after me that time. As soon as I got up I climbed back in the saddle, he stood there for a moment or two and started walking very shakily and he settled down and found out what I wanted him to do, and he never threw me or anyone after that, and actually became a very good saddle horse. As he was a stallion, he bred a couple of mares but his colts didn't suit us so we had him castrated as we wanted foals that were a little smaller and more refined. Our other stallion Chief was later used as a show horse along with some of his foals and he took many trophies such as champion stallion and champion get of sire, along with showing three of his offspring which we were riding. One time at the beginning of the Cloverdale Rodeo in B.C., the opening ceremonies had the R.C.M.P. musical ride parade in front of the grandstand; Merelie and I were asked to lead them, each

riding a black and white pinto. Merelie rode Chief and I rode a 3 year old which was out of a thoroughbred mare and Chief, a beautiful black and white tall pinto mare. She hadn't been broke and trained to do this parade work so she was very skittish when the people started clapping, she never bucked but was certainly stepping lively, and I knew I was very pleased to get that over with before she blew up or bucked, of course Chief liked crowds and really strutted past and did his part of the show very well. We used Chief for leading out trail rides part of the time and used him for cattle roundups in the fall, and sometimes would be out all day in 2 feet of snow but he was tough and never played out. One time there were several ranchers gathered at a certain location on the range and we worked all day chasing cows out of brush and from under trees where they had gathered to get out of the snow. We were up at about 4000 feet and the snow came in early and a lot of cows were still out on the range, it was a lot of tough work all day, and when we had all the cattle moving down to lower ground and started home it was getting dark. One of my neighbors rode an Arab stallion and he and I were home before the rest got down who were riding quarter horses, and we were really pleased with our horses, Chief was ½ Arab also. Another time Merelie and I had a rough days riding with Chief and a half Morgan mare we owned named Flash. We used to let the horses graze in the fall on the side hills in the lower parts around the ranch, usually the horses would come down to the ranch once the snow started to fall. This particular fall, the snow fell early and the horses didn't come down to the ranch, they hunched up under the trees for the night then would paw and graze during the day. We always kept 2 or 3 horses at the ranch just in case we needed to do something like this. Anyway after waiting a few days, we decided to go out to bring them in, and this particular fall there were a number of long yearlings with the older horses and I think they were probably the ones that didn't want to come down as long as they could find something to eat. One morning when it looked like a nice day for riding, we ventured forth thinking this would take a couple of hours. To start with, we couldn't find the horses, we spent at least 4 hours looking, and usually Chief our stallion would sniff them out, but the weather changed and a heavy fog or low clouds move in with a snowy mist, you couldn't see very far or smell anything. We should have come back to the ranch and gone up another day, however we carried on in spite of being cold and wet. We finally found the horses higher up than we had anticipated, then started to herd them down, now it was starting to get dusk, however we knew the area well. We had the horses within a mile from the barn when one of the one year old colts broke away from the others and started back up the mountain, and I was lucky to have been riding the stallion, so went back for him and Merelie followed the rest of the horses down. That colt couldn't get away from Chief and I finally got around him and started back down and I know Chief was just as anxious to get home as much as I was for he gave that colt a run and I got down maybe ½ hour behind Merelie. After feeding and looking after the horses we had been riding, and also putting out feed for the rest of the animals, we certainly enjoyed a warm soak in the tub and a hot meal, and hoped we would never have to do that again.

We had one quarter horse mare who was what we call a real fox, she was smart, each year she had a foal, and although they were weaned by five months, when the horses were out grazing she would have all her family with her , the spring colt, the 2 year old, and even the 3 year old if we still had it, and when they were out on the side hills around the trees, she would hide when it was time to come in for a days riding and I would have to go out with Chief and find them, as he was a stallion he would smell them and usually knew where they were. Eventually we put a cow bell on her neck, but she got wise to that and would stand real still so the bell wouldn't ring, especially every morning when we needed her for trail rides. One time this mare, named Cindy, twisted her ankle or hip and we couldn't use her for a while. When we saw she wasn't limping any more, we started to use her again. Well that was really something, as soon as a guest got on to ride she would limp again and of course didn't have to work. I had been watching her around the pasture and she didn't limp when not working, so we put a saddle on her again and I was there, and as soon as she started out on the trail ride, she limped again so I looked her over and tried to find any hot spots on her limbs that she limped on and I couldn't find any, so I said to our wrangler to use her on the trail, however she limped so bad the guest wouldn't ride her so I took her back in the corral, took the saddle and bridle off and let her go out in the pasture. She went off on the run happy as could be, she had fooled us again, however when I saw she wasn't limping at all, I went out and got her right then and saddled her up and put the guest back on her and that was the end of her limping as she knew she had been caught at her trickery, but she was a good saddle horse but also as I said before an old fox.

This anecdote made me think of another experience we had with another horse we called Buck, he was a quarter horse buckskin color, good saddle horse but also lazy and foxy. I do know that trail riding with a lot of dudes as we called the riding guests was rather boring as you couldn't run or do any off trail riding or half the riders would fall off their horses. One summer I think Buck just became bored to the point that he would go out on a trail ride part way then simply turn around and come back to the barn with the rider. After he did this several times and our wrangler couldn't do anything about it, especially if she had 10 or more riders to look after, so one day when Buck came back I thought I'd take him out for a ride and smarten him up, so I jumped on the saddle and went galloping around the barn from the corral and I hadn't checked the chinch band around the horse's belly that holds the saddle in place, some of the horses would hold a lot of wind when the saddle was being put on and therefore the saddle cinch would be quite loose, and this is what Buck had been doing also unknown to me, although I should have checked the cinch band before getting on the saddle, anyway with the cinch loose, it came unfastened as I went around the barn on the gallop and as I intended to follow up with the rest of the riders, the saddle and I went sailing off in mid air. You should have heard the laugher from our guests who were around there when it happened and needless to say I was really embarrassed, and I did put the saddle back on Buck properly and gave him a little workout and it never happened again with him coming back. I guess he thought it was better to go on the ride than come back to the barn.

There were lots of things that went on to cause excitement beside the horses. One time after supper, Merelie left Faye to look after the store and front end while we went out to the barn to look at the horses and just check things over, which we had to do during the busy season occasionally, this evening while we were out, Faye was serving a customer gas, and while doing this the customer was looking up over the lodge building quite intently, so Faye looked up to see what he was looking at and there was a chimney fire from the fireplace in the lounge, well that set off a real commotion. Faye ran and hollered "fire," called Judy who was in the lounge and said to ring the gong so we would come back and help. We only had garden hoses for fire equipment but everyone available got in the act, and we got the fire out before the roof caught on fire luckily. One guest who said he was a fire fighter grabbed an axe and was going to start cutting a hole in the roof to see where the fire was coming from, and I think he was nuts as you could see it was the chimney and the fire was shooting up about 15 feet, but we watered the fireplace down in the lounge and it settled down after burning out the soot build-up in the chimney. We had a large fireplace in the lounge, it would hold 3 large logs 6 feet long and 18 to 20 inches in diameter and was really comfortable in the evenings when it started cooling down, and Faye and Judy had been enjoying the fireplace when the fire started in the fireplace.

We had to go to Vancouver quite often to pick up supplies for the store and café. One time Merelie and I both went and left Faye to look after the front end and the store. We arrived home late and Faye had a very busy day and had sold quite a lot of gasoline that day, we has Esso pumps, however Faye collected some items from two different customers who didn't have enough money to pay for all the gas, she had a man's wrist watch from one fellow and a good tire and wheel from another, but they both came back later and paid their bills and collected their items. She said they were not getting away without some security.

At the opposite end of Allison Lake was a government campsite, and some of the people stopping there used to leave garbage around and of course it attracted bears, but in the fall when there was no more garbage, the bears would start looking somewhere else for food and although we were very clean as to garbage, the occasional bear would show up at the resort. One day someone came in and said there was a bear out where some children were playing, so Merelie went out hoping to scare it away, but it didn't want to leave, it was a yearling cub, so eventually Merelie had to shoot it as it had no fear of anyone. Another fall I had to shoot a much older bear that was coming around at night opening the garbage cans. This particular night we had some friends in one of the cabins close to the lodge where we lived, this friend heard something outside their door about 11pm, I got up to see what it was and when she looked out the window a bear looked back in at her, he was standing up on his hind legs looking in the window. You should have heard the yell; we heard her in the lodge and thought someone was being attacked. When she yelled, the bear ran towards the lodge and when I came out, it was really dark, I almost ran into the bear, but the woman who had yelled saw the bear run and stepped outside her cabin and saw me come out and yelled at me to watch out I was running or walking towards the bear. This bear had been around the resort for about a week as I had seen garbage cans turned over but thought if there wasn't anything for him he would move on, but there were very few guests around and I didn't want to destroy him unnecessarily. As it turned out, there were several cabins with patrons in and with the yelling, almost everyone was running around in the dark and Mr. Bear wasn't going anywhere, he was more or less staying in the backyard of the lodge watching the people and sniffing the air, as I was sure someone was going to get injured as surprisingly people were getting too close to him, so I shot him and that was the end of the excitement.

When I shot him he just sat down on his haunches, so I left him there for the night. The next morning he is still sitting up dead, so I thought I'll just move him around to the front of the lodge and set him up there for a day or so. The highway ran right in front of the lodge, so I thought we will see if anyone comes in from the highway to tell us we have a bear in our front yard; I don't remember if anyone came in pertaining to the bear. We took a picture of me petting it and left it there for several days, but we never did have another bear at the lodge to my knowledge. This was about 1965 or 66.

The years went by and we of course were always endeavoring to improve our business and we had in mind to some day build more rental units and upgrade with hot pools and etc. and also a proper RV section as that was becoming quite popular, in fact we had up to 30 or 35 RV units parked along the waterfront at times. We were planning on increasing our electrical system which was supplied by water and a pelton wheel, and we also had a Caterpillar generating unit for backup power, but it was costly compared to water power. Along about this time the Princeton Light and Power decided to supply electricity to the area that we were in. We accepted their electricity temporary depending on cost, so as it turned out we never did upgrade our own electrical system.

One fall about 1966 we bought about 70 head of year and a half year old beef heifers, a nice increase to the herd we presently had. We had quite a time trying to get grazing rights, as the former owner had lost their rights because they didn't put any cattle on the range for a year or two. When we brought the new heifers home we turned them out in the pasture; however the cattle we presently owned had not come down yet. We thought these new ones would stick around as there was lots to eat for them; the one side of the pasture was not fenced and went directly to the range up a steep hillside. One morning when we arose, all our heifers were gone, and when checking for tracks, we found they had taken a horse trail to the opposite end of the lake and open range. We were really concerned as there was no brands on any of these and there was a rancher on the opposite side of our range and valley that had a long rope and he made no bones about it, and said that any animal that had no brand and he caught, he kept. We waited for a few days as we thought they would come back as there was much better feed in the pasture, however they didn't return so one of our employees and I drove down to the opposite end of the lake and I walked up in the bush to where this horse trail ended, and told the employee to come behind me very slowly but keep out of sight in case the cattle were there which they were. I had seen them and was trying to get around and above them so that when they saw either of us they would follow the same trail back to the ranch. That didn't happen, he came up on them too quickly and they saw him and started right up the hillside on a deer trail, so then I ran and got on a higher deer trail and tried to frighten them down, they were on a trot and I thought they can't out run me. I was in good shape, and as it turned out we traveled about 2 miles gradually up all the way, and I kept pace with the leaders but couldn't get in front of them and eventually they stopped, looked at me sitting down by this time just in front of the leaders, and from where I was sitting I was looking own on our buildings above the pasture they had left a few days earlier. There was one thing I knew and that was whether they came down or not, I couldn't run any further, so after a good rest I went around them and started down to the ranch, as it turned out that evening they all came down and never left again. The worst thing that happened was I had a mild heart attack the following morning but luckily we had a doctor guest at the resort and he attended to me. Some years later I had to have a heart valve replaced due to this episode.

Shortly after we bought Skye Blue we had two couples come in with two RV trailers to go fishing and Merelie told them where they could park and also where not to park which was in the picnic area near the dock. Wouldn't you know it; they went right to the picnic area and set up right opposite the dock. In the evening when we would walk around and check things over Merelie saw red when she saw those two trailers and she went over and the two women were in one of the trailers, in fact one of them was one she told where to park. Merelie told her that they had to move into the RV area, there was a discussion about it as their husbands were out fishing, so Merelie said to try to call them in, however they had to move when they did come in. Their husbands came in shortly afterward and Merelie waited a while and it was getting closer to darkness, so she went over and asked them "when are you moving"? They apparently said they were not or not until morning, that did it, she went over to our large farm tractor which had a large bucket on the front, she took the tractor over to the front of one trailer and went inside and said "you had better get out of your trailer as I'm going to move them, those trailers are going right in the lake, next time you won't park here in this spot." She went out and revved up the old diesel tractor and it sounded worse than the bite so to speak. Well needless to say, they moved their trailers right away and down the road they went, and any of the kids that were around at that time would tell everyone about Mrs. Broughton going to push the trailers in the lake that parked in the picnic site. I doubt if she would have done it but she made them understand what she was talking about. It turned out that these people were from Princeton, our neighboring town and it became well known in the area; don't mess with those Skye Blue owners.

We had a mortgage on the property, with large payments every six months. We dealt with a bank in Princeton and issued the payments from the bank, and one fall payment that had been issued, the recipient claimed she had not received it so arranged to meet at the bank to have it settled. Merelie knew she had sent it out and all payments had to be certified cheques. We all met in the bank manager's office, and the manager proceeded to tell us to issue another payment. Merelie said we had issued a certified cheque and were not paying another one if it wasn't certified maybe but not when it was certified. The manager who was a good friend of the previous owner insisted we issue another cheque, so Merelie said "unless you change your mind we will close out all our accounts. I guess he thought we wouldn't do that as there was no other bank in town, but we were in between Princeton and Merritt and they had two banks. Merelie walked out of the office, went to one of the tellers and said "I'm closing every account we have here," there were several and we did do quite a lot of business there. The bank manager really was apologizing and said he knew we were right, and Merelie told him to forget it, he had his chance and that was it. Needless to say we banked in Merritt from then on; however there was another part that came out of this, firstly the cheque was found about six months later behind the recipient's fridge where it apparently had fallen. Secondly, we received a letter from the Princeton bank saying they would be very pleased to give us a loan to buy cattle which we had planned to get sometime previous to the cheque incident. The manager of the bank had retired and there was a new manager there, Merelie riled quite a few and never backed down once her mind was made up.

When we took over Skye Blue, it wasn't very popular locally due to the attitude of the previous owner. We were told by her not to have anything to do with the local people, however we didn't pay too much attention to this as we planned on living in the area maybe for the rest of our lives, and we always had good neighbors.

Gradually people would come in to the restaurant and maybe have a chat if Merelie or I were around. What we finally found out there was a lot of local people coming in to check us out and they kept on coming as long as we were there.

We became members of the Princeton Chamber of Commerce, then one or two years I was president of the Chamber. We got quite a few members, even had some of the ranchers take part, and anyone who sold anything and did any type of business was in if we could get them, as the Chamber became well known. It really helped our business as when some stopped in Princeton looking for places to go, they always sent them to Skye Blue, our tourist business really grew. When we became involved in the Chamber we found there was government grant money available to promote tourism in the province and up to now Penticton Chamber had been taking it all. Our Chamber got busy and organized the whole area of the South Okanogan such as Merritt, Hedley, Osoyoos, Oliver, and Ok Falls. We wanted our share of the grant which we got, and no doubt everyone prospered from this.

We had some type of activity going on every holiday or anything that we could codger up. We had dignitaries from various newspapers from as far away as California, and government dignitaries and members of Parliament. One year we had the old stagecoach from the tourist bureau with horses and drivers and outriders going through Princeton when some government officials were there, so the stage gave them a ride around town; however what they didn't know was there were some masked riders that were going to hold up the stage, which we did. The robbers took everything of any value and left. I guess it shook some of them up, however when the stage got back to the tourist bureau, there were the robbers with their things. That group never forgot Princeton. We did all types of things as I said before with parades, horse racing, and everything for excitement, everyone in the South Okanogan took part and took some time off for playing. I remember one parade for Princeton one of our employees and her mother who was a librarian for Merritt made up a covered wagon. Merelie and I were dressed up as the pioneers and were with a team of Palomino horses. We had a milk cow tied on behind, the librarian had long hair so braided like the early Indians did, had a buckskin outfit and rode bareback with moccasins on her feet riding on one of our red and white pinto horses, and on each side we had a scout riding on pintos. There were a couple of instances that happened while showing this setup, our horses had never seen white lines as they hadn't been used anywhere but on the ranch or resort. As the horses approached the white lines at crosswalks, each horse would stop and almost look both ways before crossing, and you would hear the spectators say "look at those horses, have they ever got those horses trained, they stop to see if anyone is in the crosswalk," which we of course got a kick out of their comments. I had told the rider on our black and white stallion to watch him as he had never been in a parade and may act up with clapping etc. As it turned out Chief, our stallion was the proudest one in the parade with his head up high and stepping like a real show horse and when we got to the end, he kicked his heels up in the air and gave a squeal as if to say "I showed my stuff." The rider said to me after "you were pulling my leg about that stallion," however I assured him I didn't know what he would do, although after thinking about it, his sire was a show horse in Victoria for quite a few years and was well known.

One year the Princeton Chamber of Commerce had a beard growing contest and I was president that year. I looked more like Gabby Hayes of the movie set than he did, I believe it was 1966. We would also have one of our vehicles prettied up with the Chamber of Commerce sign on to advertise our area.

We supported the Social Credit Provincial Government at that time and knew a number of the members of parliament. Some years later I was asked to run for election in our riding but turned the offer down as we planned on selling the resort and ranch and planned on moving out of the area.

Shortly after buying this property, in fact the first winter, Merelie and I made a beautiful sign to put up in front of the lodge, as highway 5A ran right through our valley in front of the lodge. The sign was made out of 4 inch thick, 4 foot wide and 8 feet long pine planked with chiseled inset name and hung on an 8 inch pine log which sat on two 8 inch posts, white letters with oiled background on the plank. We also had an old logging arch with the 12 foot wheels next to the sign, really caught everyone's eye, and there were a lot of pictures taken of this. There were only two in B.C. that I am aware of.

When I think of this logging arch, it brings back memories of some of our plans. There was the old original log house on the property when we bought Skye Blue, which we had big plans for, we were going to fix it up and have an antique display. We started gathering up old items from around the area. When we sold, the buildings were full of relics with several larger items outside, like an old buggy, wagon, a sleigh that belonged to Jack Bud who had been a friend of the train robber, Bill Miner. When we sold, we wanted to hold these items back and reduce the selling price $15,000.00, but the buyer said "no" he would like to make a show place as we had planned. As it turned out he destroyed or buried most of the stuff and also the old house.

This property had been homesteaded by a man by the name of Burns in 1889, we were the third owner. The lake which is known as Allison now was originally named Burns after the old homesteader Burns.

Due to over work, not taking time out to rest, Merelie had a partial stroke in 1966 during the summer. We were fortunate at the time to have had some of our family fill in and help out at the busiest time of the year. She was in the hospital a short time and gradually got over it, caused by fatigue. I had grown up on a farm and learned that your work was never done, however Merelie couldn't bring herself to leave work as it would be there tomorrow as it was every day in a resort, ranch, or farm, they were all the same. We thank the Lord that her health returned for another year.

In 1967 Merelie had an aneurysm burst in her head and was in the hospital again for a couple of weeks, this time she realized she had to slow down or she wouldn't be around to enjoy the scenery or work. We were fortunate this time as we had a young doctor that had been studying about aneurysms for the past two years and he tapped Merelie's spine and drained the blood off her brain, as the aneurysm is a blood vessel that bursts then leaks, and in this case into the brain. This operation made medical history as it had never been tried on an animal or human. Merelie was in the hospital unconscious for 10 days. Her hair turned white from a dark blonde and lost part of her eyesight and her memory was lost temporary. On the tenth day, she woke up and asked the nurse what she was doing there. We had a nurse with her 24 hours a day, and the nurse who Merelie knew, said "it's not what I'm doing here but what are you doing here"? Merelie didn't remember anything from before she left the resort to then. One day she had just come into the kitchen, sat on a chair and said she had the worst headache and was holding her head in her hands. Our son Keith drove his mother and me to the hospital in the shortest time possible. As soon as Merelie saw me, she said "don't tell anyone I had an aneurysm, they will think I'm crazy or something," and she also told all the family the same "don't tell anyone." It took a little longer to recuperate this time and we hired more help as this time she realized she couldn't keep up the pace she had been doing, and everything went along as usual, however the writing was on the wall as the saying goes. We would have to sell or Merelie would not be around much longer, and now she started worrying about my health, which wasn't necessary. I was enjoying myself as I knew I couldn't do all the work in one day.

Hindsight is a very good thing if a person could foresee the future. We should have had Faye and Keith come back to take part in the affairs. We thought about it but didn't say anything to them as we had in mind that they would rather be doing their own things. Keith had a steady job and so did Faye. She worked in the lower mainland in the Vancouver area and Keith worked in Princeton and then in the Merritt area. As it turned out, later they both said "why didn't you talk to us?" They liked the place only I think maybe I was too ornery and although they had both worked at the ranch and resort for a couple of years, had eventually gone on their own. There was no doubt the place had a good future and lots of room for improvement.

The ranch was bringing in a return of about 7 ½ % return on the invested dollar, which I must say was much better than most ranchers, in fact while we were there, the government had a financial survey of all the ranchers in our area and the highest return that was recorded was 2 ½ % on a ranch in the Princeton area and that was Wayne Sellers. The resort was doing very well, as we also had some winter business and was always booked up during the Christmas and New Year season. We had plans for expansion, almost had our mortgage paid off, which when done, would have given us more working capitol. We always had to have a large Christmas tree and everyone had fun getting a tree. We would take a team of horses and a sloop and go up in the bush where there were some nice trees and pick out a nice tree, bring it back, put it up, and decorate the tree and lounge for the festive season. One year some friends came for a few days and they put on a little skit and dressed up as reindeer on Christmas eve and everyone had a great time.

The horses were classed as part of the resort as they were mostly all used for trail rides. We did have a number of mares and a stallion which produced 5 or 6 foals every spring. Some we kept for trail rides and some were sold. Two went to a baseball player in Winnipeg. We would sell a foal for $175.00 to be picked up in the fall, that was much better than the cattle prices.

Another winter my brother and his family came for Christmas and had a lot of fun playing in the snow. At that time I had cleared some trees off a side hill for a little skiing, and there was a ditch running across the hill about 20 yards from the bottom of the hill which gave the skier a good place to make a jump if you wished and sometimes even when you didn't wish. This one year there was a lot of snow and we were out playing and skiing down the hill. My brother at that time smoked a pipe, and he skied down the hill, hit the jump, went up in the air and came down head first and broke his pipe off which he had in his mouth. Well everyone had a big laugh, he said the hill was so short he didn't think about the pipe. I'm the next coming down and wouldn't you know, after laughing at Lorne taking a header in the snow, I went much higher and deeper, I turned a summersault and had snow packed down inside my shirt, no one could have had more snow in their clothes than I had, and of course everyone really laughed at my fancy landing on my head.

One fall we had a truck load of cattle to sell and we had them in the corral loading them. It is seldom Hereford cattle are dangerous to handle or move about, however we did have one that was crossed with another breed, a steer with long horns. Merelie was in the corral chasing the cattle up the ramp and into the truck, but this one steer apparently didn't want to be loaded, I saw him take a run at Merelie when she had her back turned and told her "look out, you had better get out of there," I was sure he was trying to gore her with his horns, and she said "I'll watch him, I'm ok," and all of a sudden he took a run at her but this time luckily she saw him coming and she climbed up the side of the corral, he kept coming and hit a log in the corral with a horn right below her feet, and knocked a piece of wood out of the log. She was fortunate to get away from him. He then carried on up the ramp, and instead of going in the truck, he jumped over the ramp fence, fell back down in the corral then went through the rail gates, out into our pasture and kept on going up a side hill and out towards the open range in the timber. We didn't bother with him, finished loading our cattle. I figured when it got cold enough and the snow deep enough, he would come back. Later that fall one day when feeding the cattle, I saw him standing near the bush on the far side of our pasture. He eventually got hungry and brave enough to come up to the feed racks and eat with the rest of the stock, but when he saw me, he would go on the run back to the trees. I kept watching him and planned on eventually using him for meat but had to get him when he was quieted down or the meat would be tough. I always kept a 30 30 rifle in the barn just in case I needed it. Well one morning in mid winter we needed some meat and there was Mr. Steer standing behind a feed rack with his head sticking out. I just took the rifle and shot him in the head through a large knot hole in a board in the barn. He didn't have a chance to run that morning, so I took a tractor with a bucket, picked him up and butchered him after I finished feeding the cattle. We still have his horns which are over a 2 foot spread. The meat was good with very little fat, and there was one thing for certain, he wasn't a Hereford although he did have the coloring.

We always branded our cattle before they went out on the range, this we did in a cattle shoot as we thought it was much easier than roping and tying them as most ranchers did. One time we were doing this and I almost lost a finger. We run the animals in the shoot and stick a heavy bar through the shoot behind each animal into a hole. This day the animal came into the shoot and started to back out and I put one hand behind the cow and unknowingly put one finger in the bar hole, and Merelie pushed the bar through and took all the skin off my finger. I'll say two things, it was sore and I will watch where my fingers are.

The Morgan mare we had would go off and hide to have her foal and would bring it in the barn when it was strong enough to travel. One time she had a foal out in the edge of the field in some bush, and apparently the foal must have started to jump around as it was injured quite badly, looked like it had jumped on a jagged stump where we had been cutting timber. Flash, the mare, brought her foal in and started whinnying at the corral fence, she made enough noise that we went to see why the noise. The foal needed some immediate veterinary attention; a lot of his belly was torn up and required a lot of stitches. We did the work ourselves and didn't expect he would live, however he did and was a tough little horse, so we called him stumpy, after the stump he had tried to jump over.

One year Merelie bought a 4 month old Hereford calf from a neighbor who had planned on destroying it as it had been mauled by a bear, it was in quite bad shape but Merelie brought it home and doctored it up and it started to heal, well that calf became a pet for quite a while. Merelie would go out and call it and it would come on the run and would lie down while she washed the wounds and put in healing disinfectant, in fact she used a veterinary product called Scarlet Oil. This calf had been clawed quite seriously on both sides and back and also had chunks of its hips missing where the bear had bitten it out. That calf healed very well and when it was about two, you would not know it had been injured. This product Scarlet Oil was like a wonder healer and we always had a colt that needed to be doctored up, they would try to jump through a gate or over something especially when their mother was taken away for an hour or so on trail rides. This product would heal them up and never leave a scar. I had used it on my hand one time when I had a very bad injury and my hand healed up without a blemish.

One fall while we were cleaning up at the ranch, Keith was driving a gas tractor that we had borrowed because our diesel tractor was away for repairs. He was cleaning out the large barn and was in the barn too long with the tractor running and was almost overcome by carbon monoxide, and as he drove outside, he fell unconscious off the tractor and unknowingly shut the tractor off and landed between the front and rear wheels. We picked him up and took him outside the corral to give him some fresh air, and he said he wondered why someone was slapping him and asking what his name was, and he gradually got his senses back. That was a close call and we were much more careful after that.

Something also about Keith comes to mind; one spring we bought a nice looking small gray mare from some people in Princeton. We thought she would be real good for the guests to ride, but we made a real mistake that time, that mare could buck and throw riders off faster than we could put them on. As it turned out, the family that had her also had two boys who enjoyed her and she thought that was what she should do, and I'll tell you, she could dump anybody, even the best riders. As it turned out we found there were certain people she wouldn't buck, one was a teenager we had looking after guests on trail rides and also a young woman that we hired for trail rides and she would go for quite a while without bucking someone off. One time after we thought she had settled down I took her out with some guests to look at the ranch, and as we were riding along smoothly, all of a sudden I was sitting on the ground, she didn't run away but came over to me to ride again which I did. It appears to me that she did a little sachet, a slight twist and walked out from under the rider, but she never bucked me off again but I never trusted her either. Later on that season, Keith wanted to go riding with some friends and the only horse left to ride was the gray mare, so Keith took her and Merelie said that if she bucks you off, make sure to get back on. They were riding out on the ranch when a grouse flew up and Keith landed on his head and bit his tongue, he got back on her about six times but each time she would buck him off, and when they got back, one of the riders said "Keith is sure a bear for punishment, he was bucked off so many times." We finally sold that mare to our female wrangler who was never bucked by the mare.

A lot of the friends we made at the resort would come for a week and some would stay occasionally all summer and some off and on all summer. Most of our guests came back every year, and we had two groups that came for a week each, every winter thy would rent everything available, this I believe was for 7 or 8 years until we sold, and they probably continued after that.

When we sold the resort and ranch in October 1969, the new owners took possession Jan. 1 1970, and they were not the most hospitable people, it was two couples and they agreed to disagree in the first year and it was resold in one year. The people that were buying the property thought all they had to do was sit around and drink coffee and talk to the guests or whatever. This place went through five owners in five years and went down hill until the last owner set fire to the lodge for insurance as he owed $155,000.00 at that time and thought he would clear his debt. None of the new owners were ranchers; there was no upkeep on anything. The first owner put in a liquor bar in the lounge and I don't think it helped the business any. We had been paid out by the third owner who was a friend of ours who had bought it for his son and daughter-in-law. They had known that it was very profitable when we owned it, and only the son didn't realize there was a lot of work to it.

When we were owners we had become very active in the Motel and Resorts Association where I had been vice president. We had some good conventions and did a lot of advertising to promote B.C. tourism. We even had a tourist booth in Seattle which Merelie managed for several seasons. We would have these gatherings and promotions during the fall when things were a little slower and we had time to get away. The owners that followed didn't take part in any of these activities to my knowledge.

When the lodge burnt, the property was purchased to develop into lots along the water front and the ranch went into almost waste land, a waste of property to my estimation as it was an ideal resort location with proper management. We tried to buy it back as the local credit union was holding the mortgage, and we would have paid them out in full but there was some deal made by someone else with the manager so we didn't get it. We had time to recoup and get our health back and had some new ideas to develop a resort, it would have been a deluxe operation and we would have been worth well over a million dollars today if we had got it back.

When we sold, the buyers didn't have much cash so we took over a 21 suite apartment block on Kitsilano beach in Vancouver, also a house and office block in Burnaby. We moved into the house, paid out the only shareholders we had, Mick and Ted Johnson, who we sold shares to as working partners along about 1966, they stayed for two years and then they left as Ted couldn't get in enough fishing and hunting, said we would pay them out if we sold, and they would get interest on their investment. When we paid them out, they had made a good return on their investment. After settling in at Burnaby, we went out and bought a new car, a Mercury Marquis Brougham, a real nice car.

While we were at Skye Blue during the Christmas season of 1959 Merelie and I had a real bout with flu and colds, I guess we were run down and it took a while to recuperate, and finally we felt well enough to travel and we went south, through California, Nogales Arizona and down to Guadalajara Mexico. We had a good holiday, drove around quite a lot and saw the area, especially Lake Chapala which is quite a lot like the Okanogan of B.C. We had an experience with Mexican traffic. One day we were coming back to our motel, in fact only ½ block from it, we stopped beside a bus while they let out passengers, but they don't drive over to the curb as our buses do, they stop in whatever lane they are in and everyone is supposed to stop to let the passengers go to the sidewalk, however a cyclist came peddling through and ran into our bumper, bent his front wheel and fell off, put up a big holler that we were wrong, anyway the end result was we gave him a few dollars, he went away happy carrying his bike, otherwise we would have been held up for several days with the Mexican adjuster and all the red tape. Fortunately we met a young American who lived in Guadalajara and he pointed out the easy way to handle it.

Everyone was very friendly and I learned enough Spanish in a couple of weeks to get by. We found that the Mexicans are real good workers, the most of them take an hour or two off about 2pm called a siesta, but they start work very early in the morning and are very happy people all the time. Whenever we were touring around, there was always someone willing to give directions and help if they could. One time we went into the town of Chilacipaque, a boy came along and wanted to be our guide, and he also wanted to learn the English language, and some day he planned on being a full fledged guide for tourists. He showed us a lot of places, were right into people's homes and some of the little factories where they were doing crafts and making things to sell to stores. We picked up some souvenirs of course and had a real good day. When we wanted to pay him he would only take 50c or something like that as he said he couldn't take more because he couldn't embarrass his father by earning more than him, he did want some pencils and pens and would like to have an English dictionary which we bought and sent to him when we got back home.

The Mexicans drive quite fast for the roads they have, which are narrow. They had frightened us a number of times, however we did have one up on a pair of speeders in an Iseta two seater on our way out of Guadalajara. There were 4 or 6 lanes going each way, and we got crowded into the turning lane, but I had no intention to turn, so I proceeded straight through toward the traffic in the opposite turning lane with the blinkers indicating that I'm moving over to my right, but the Iseta with 2 Mexicans in it saw our big car and thought they were going to be wiped out. As we approached them, the driver threw his hands in the air and by the mouth movement he hollered "mama mia," however we do not know what happened to them, we passed them doing 45 miles an hour as that is the city traffic speed and we had no time to see how they made out but there were 2 frightened Mexicans.

When we sold the resort I was planning on retiring. I was 52 and Merelie was 47 and we thought we could take things a little easier, however time will tell.

In the spring of 1970 the couple that was managing our apartment block wanted to retire so I decided to sell it. I went to a real estate friend and told him about it and he asked me what I was going to do, so I told him I didn't know but I couldn't stand retirement very long, and he said "how about selling real estate," and I said "why not, when do I start?" so he said "you have to get a license first," so I went to U.B.C. for a short course to learn the law and regulations, then I wrote an exam and took another short course in one of the real estate offices to learn the tricks of the trade, then I had my license. I now started selling real estate for my friend on 4th Ave. in Vancouver. The first thing I sold was the apartment block, of course I had been selling and buying real estate for my wife and I for quite a few years, so this wasn't any different only now I had a license to sell someone else's property.

The apartment block had 21 suites, and I think I sold it for $250,000.00 with a first mortgage and an office block on 8th Avenue in Vancouver, which had several mortgages on it, kind of a tricky deal but in those days it was fun. I eventually sold the office block to a real estate salesman from a downtown company who thought he was a real cracker jack. He had in mind something that didn't work out for him. He made mortgage payments for a while then came to me and wanted a discount on a payout which I wasn't prepared to give him, number one lesson in real estate; anyway, he said "you will wish you gave me a discount because I'm going to miss making mortgage payments." At this time I told him to read his contract over thoroughly as things may not work as he wished. I then contacted the other mortgage holders; I think there were 5 including his first. He hadn't been making payments to them and had been pressuring each of them for a discount or delayed payment. I paid each one of these mortgages up to date and didn't say anything, and I also told them not to tell him I was paying them. Well the end result was he became 3 payments short over a six month period then I turned it over to a good lawyer for foreclosure. When he heard from my lawyer he was real upset, and it finally went to court, and he pleaded that he didn't understand the law, but the judge didn't buy that and gave him 3 months to settle all the mortgages or the block would be resold, and as it turned out he got someone else involved and we each received our money with interest.

I sold real estate that summer, and as I liked selling, I had fun at it. In the fall we went south to Arizona and California. In the meantime Merelie and I were looking around for some acreage in the rural areas. We had kept several horses when we sold Skye Blue and needed a place to keep them and do a little hobby farming.

Sometime after we returned home from Mexico Merelie went out and bought a 25 foot Kustom Coach R.V. trailer against my wishes, as I had seen some of the problems people were having with campers and trucks and didn't think the trailer was much different; however she did have some good arguments for buying the trailer. We would have our own bed and eat as we wished and stop where we wanted. We had a hitch put on the car and brought the trailer home which was two blocks from the sales yard, and what do you know, I put a hole in the trailer at the neighbor's garage.

In the spring of 1971 I started selling real estate again, and got involved with a group of real estate salesmen and bought a real estate company, however shortly after this deal I found that several of these salesmen were sharpies or didn't mind twisting the real estate rules so I got out of that deal about $5,000.00 short, but better out than loose my license.

The summer of 1971 Merelie and I went to the east coast with the trailer for 3 months. When we returned we bought 21 acres on 66A

Ave. near Cloverdale and built a house there where we lived until 1986. We eventually sold our house and office block in Burnaby. We went south again to spend the winter in Arizona and California in 1972 and came home in the spring. I started selling real estate again for a friend in Surrey, who also belonged to a horse club which we belonged to. We had a lot of fun showing horses some of which we brought from the ranch near Princeton.

Real estate was really on the move and I was doing really well, and had an experience one day where 2 men stood in front of a piece of property I had for sale and argued about who saw it first. The end result was I told them that whoever got in the office first with money would have it, so off thy went, and all I remember is that one of them bought, and the other looked for something else.

The winter of 72/73 we wandered around California and Arizona, visited some cousins in Pomona, also some in Medford Oregon. When we returned to Canada we applied for U.S. citizenship and were received with open arms, now we had the ability to do business in the U.S.A. and reside there. The spring of /73 we sold the trailer and bought a used Barth motor home, and also went back to selling real estate for the full summer then back to the U.S. for the winter again.

In the spring of 1974 I became involved in placer mining, had a fellow looking after our stock on our little farm in Cloverdale while we spent our winters south. Anyway he had been placer mining for quite a few years, hadn't made any money but had a patent on what he called a gold panning machine. Didn't have any money to build it but had the knowledge, so I said I would supply the money and we will build the machine, find some gold producing property and see if it works. The end result was we built the machine which cost about $4,500.00 and then we found a property to put it on, which was up on the Little Joe Creek off the Fraser River near Big Bear Creel on the opposite side of the Fraser. We had to take the equipment over by raft then tow it up to the claim on Little Joe Creek. We didn't own the claims so agreed to rent the machine out to see if it worked. Well it did but we were not getting paid so we moved it off the property onto another one near Lillooet on the Fraser River. The three claims which we had checked out, according to the geologist would run about $100.00 per yard in the ground. We didn't own these, we had a better deal, and there was the claims owner, the fellow who was financing the operation and

ourselves. We would split whatever came out of the ground 3 ways or the other two could buy the machine and we would build another. The machine really produced the gold, we had a trial run of 5 hours and the pan that caught the gold was 5 feet wide and 14 feet long, and there was a strip of pure gold dust down the pan one foot wide and 5 feet long. Well I can tell everyone was happy. Now it had been arranged with a company in Wenatchee that they test what we had and said it was running about 90 ounces to the ton of gravel, everyone was about to become rich, however the fellow who took the samples was paid for what he had and proceeded to get drunk and he never sobered up until he killed himself and someone else that was with him in an automobile accident. Before he killed himself he had been on a big drunk that lasted about two weeks while the rest were waiting for his return. When he was no longer in the picture, his wife said she didn't want anything to do with it, and besides he and she didn't need the money as he was very wealthy and just wanted something to play around with.

Now we said we know this is a worthwhile project, the claims owner and ourselves would carry on and split 2 ways, however the claims owner now wanted it all but had no money to operate, and there was no way we could get that claim owner to agree to anything. Eventually we filed a lien against the claims for our expenses which were at that time $37,000.00 and we moved the machine off the property. He could not get anyone to go in and work the property at his terms and eventually sold those claims for $25,000.00, the buyer bought them for a future investment. We had a number of men with claims come to us but we had our share of placer mining and eventually sold the machine for $14,000.00, not very profitable but a good education. You wouldn't believe the type of people that tried to get that machine but no one had the money to operate or pay for it. This placer gold mining venture covered a period of 2 summers or the better part of them.

In the summer of 1974 Merelie and I made a trip to Cassiar B.C. as our daughter Faye was working at the asbestos mine as a lab technician. It was quite a trip, the highway up there was not the best, although we only had one flat tire, right in a real muddy area too. That is a very beautiful and interesting area of B.C. Later that year Keith and I went up there with a truck to move Faye to Surrey. We stopped at a restaurant on the Alaska Highway one morning for breakfast and ordered hotcakes, bacon and eggs, the waitress asked how many hotcakes, so I said three, so when we got them they were ½ inch thick and the size of a large dinner plate , well I could only eat one. This was in the fall when a lot of hunters were going up there, and I said to the waitress "I didn't know the cakes were that size," so she said "eat them up, you will need them before you get back."

While still selling real estate, I had a rancher come to me as he wanted to sell his ranch which was about 20 miles west of Quesnel on the west side of the Fraser River, so I took a listing from him. The ranch comprised of 95,000 acres, buildings, and equipment and he owed $25,000.00 on it. I asked how long he had been there and his age, this was a large ranch for B.C. He was 67 years old and had come up to B.C. 30years earlier, had been ranching in Montana and decided to move up to B.C. and bought a small ranch, then went back and drove 600 head of cattle up to his new ranch. Over the years ranches adjoining his came up for sale and he bought them out so now he owned all this land and was getting too old to operate it properly. It was a beautiful piece of land, almost all flat, nice small lakes and creeks through it. He had sold all his cattle and now wanted to retire, his wife had passed away, his only child, a son was in Alaska and didn't want to ranch. I did have an American group who were looking for a ranch that they wanted to ranch and run a tourist operation together, they offered to put down $50,000.00 with an option to buy or pay for the ranch in 3 years, they would pay him a salary to look after the ranch, buy more cattle and they would proceed to develop a tourist operation, and if at the end of 3 years it didn't work out tourist wise, they would walk away and he could have any improvements they had done, but he turned it down. Within the year the ranch was sold for a small down payment as it was on multiple listing by another salesman and the rancher died a year later, of course after that I had no interest in the ranch so I do not know what is going on, however because of the size of acreage I may go there some day just for curiosity sake.

Whenever I wasn't involved with real estate sales and working our little 20 acre farm, we had a few cattle and several horses which we had for the horse club for play days and horse shows. One year we showed three pintos and the stallion as get of sire, and won a number of trophies and almost took all the prizes for different events. Well that is not the thing to do if you want to be popular with your club members, anyway it was fun.

Every once in a while we would get itchy feet and would go for a trip in Canada. One summer Merelie and I with our car and trailer and my brother Lorne and his wife Phyl went to Thunder Bay Ontario through the northern states to Bemidji Minnesota then up to Canada. When we were traveling, we would stop for breakfast or lunch along the road. As they had a retired school bus made into a motor home, they would always be ready to leave long before we were as we couldn't prepare anything until we were parked and this started a new idea, for when we got back home maybe we will sell the trailer and buy a motor home. We had a good holiday and visit with relatives and friends. When we arrived home we sold our trailer and bought a used Barth motor home which we used for a while, it needed some spit and polish. We went back to the dealer with it and he thought it was a different one. The following year we were going to purchase a later model, but went looking at RV's in California. We ordered one, a new make on the market, a1977 King's Highway, manufactured to your specifications, 31 feet, gas model with a 440 Dodge motor. We came back to Cloverdale and sold our Barth motor home.

We had always tried to keep a farm status on our 21 acres, however one fall we sold our cattle which were a new breed of cattle to the area which were called Queinina originally from Italy. They were tall white animals with black on the ends of their tails, on the tips of the ears, and black noses, which were quite docile if you didn't get them excited, and once they accepted the area you put them in they would stay I believe without fences, but if they got excited or didn't like the area they would jump a 6 foot fence, and they were big animals. To the top of the hump on an average was 5 feet 9 or 10 inches. One cow and calf into a new pasture, I caught the calf and proceeded to take it to the new area, the cow jumped over the coral fence which was 6 feet and came and looked at me and mooed and wanted me to give her calf back to her, she was not vicious or tried to attack me, just looked at me and mooed, very gentle animals. The government had brought some of these cattle in from Italy, they were used for the meat, the milk and also as a work animal if necessary. The reason for having them was to try and cross breed with the rancher's beef as these cattle had no fat on their body, the more they ate the larger they would grow but no fat. As it turned out they were not suitable for ranching, they had to be handled gently or they would jump the corals or fences and leave, and it wasn't feasible for ranching. One Texas rancher took some and crossed them with Angus and he registered a new breed called Kangus.

Well back to the farm, we had to make so much money per year to qualify as a farm, so we decided to buy some brood sows. We bought one but wouldn't accept a boar and she traveled around with the cows. We would feed her in her pen, then she would jump over her fence and go and eat apples or sleep or stay with the cattle. We eventually butchered her, was exceptionally good pork and had an apple flavor. We did buy another brood sow which was going to have piglets shortly, she did but had milk fever and was going to kill her litter, and I believe she did kill one before we moved them. We had to get rid of her and raised the litter ourselves, or I should say Merelie raised them. I believe there were 12, and Merelie taught each one to drink out of a small trough, one

little male was very dumb or stubborn. I think it took Merelie almost a day to make him drink. If pigs are given the right type of quarters to live in they are very clean, these piglets had the best and became real dependant on Merelie and would follow her anywhere and do what she wanted them to do, even at the stockyard where we sold them the fall of 1977 to make up our farm quota. At the stockyard they wanted them in 3 pens for selling, she showed them where to go and almost cried to see her pets leave.

In the fall after selling our pigs, we packed up our King's Highway motor home and headed south to California. We went out on the desert in the Imperial Valley which is just north of the Mexican border. We were parked in an area that had been the army training ground for George Patton's troops before he took his trainees to Africa to contend with Hitler in world war two. This parcel of ground had been 1700 acres, but after the war the military department decided to sell or dispose of the property so they offered it to the town of Niland which was really close to the town, about ½ mile, for $1.00 and the town could do as they pleased with it. There was a complete camp there with water, sewer, electricity, all the buildings, streets, and a nice setting in the desert. The town officials turned this down, so the army dismantled all the buildings and left it. One California senator bought 960 acres or the better part of the 1700 acres and didn't do anything with it, so the snowbirds who traveled south in the winter started parking on the cement slabs that remained after the buildings were demolished. The area soon adopted the name Slab City and this is where we parked also, no rent and not much of anything else but sunshine and RV's all over the area.

The senator that had bought part of the army camp had died and his part, 960 acres was for sale by his heirs. I looked the ground over and decided to make an offer, part of it would make an ideal tourist mecca and the balance for farming. At that time a company from Texas was looking for someone to grow the guar bean, which at that time mostly came from India and that area of the world, apparently the Imperial valley had the same climate, but hadn't been grown there, then secondly there was one hot well and several more hot spots where one could drill and get hot water which was ideal for the snowbirds for winter rendezvous. The guar bean is used for children's play putty, the plastic parts of automobiles and many other items as well as fast food products such as food filler in ice cream. You never know what you are eating do you?

I eventually contacted the family and they told me to see their lawyer who had an office in the Imperial Valley at Brawley. I went in and talked about my interest in the estate and he told me the highest offer they had was $75,000.00 and he thought that was all it was worth, so I said "I'll give them $80,000.00 and he assured me I would have it as it had been dormant for several years. We waited for him to contact the owners, there was eleven involved, and the head of the group was married to a Russian Jewess and she was the staller, the lawyer thought he had a deal for us on several occasions, but every time we would meet, she would want just a little more. We did finally get to $125,000.00 over a period of 2 years and then the hot well that was on the property burst as they had a small quake in the area, so I went down and talked to a well digger in the area and asked him what it would cost to repair the well pipe and his answer was $15,000.00 or more. The area is actually sitting on a fault line and could have a major quake; however I had been willing to gamble on that until then, as it was, we all met on the property at the well which was now running water down into a canal. I told them that my offer had reverted back to $80,000.00 or nothing. This gal I mentioned before had a real estate salesman with her and he said "forget it I'll get you more money," and fortunately for them he did. Shortly after this the Mexican Peso hit the bottom and a group of doctors from Mexico came in and bought it for $225,000.00 but it has never been developed and there it sits idle, so much for that venture.

    The winter of 1976 and 77 we were with friends of ours from Grants Pass Oregon. The Stanford University prints a paper pertaining to Stanford activities and accomplishments which comes off the press either weekly or monthly, anyway we got a copy shortly after Christmas and the headline in the paper read "we now have a cure for cancer," and the article went on to tell how it had been discovered, it was a herb, where it grew and how to collect it Apparently one of the retired professors had developed cancer and he was in the Stanford hospital, had a growth about the size of his head in his body, he was an average sized man about 165 pounds and had now lost weight and was deteriating down to about 65 pounds. The university is about 75 miles north of the Imperial Desert, so this fellow decided he was going to die anyway; the university staff had known for a time that the natives living in the area were curing people with cancer, so he decided to go and live with some of the native people to see if they could cure him, and eighteen months later he returned to the hospital where he wanted to be tested, he was back to his normal health and weight, and the tumor that had been the size of his head was like a shriveled up pea and he had no cancer, however the native treatment is supposed to last 24 months for a complete cure. We do not know if he had a recurrence of cancer or not.

This article went on to tell what was used, how to pick the herbs and how to use it. The herb grows quite prolific in the Imperial Valley desert; it looks like a brown willow brush up to 18 - 20 feet high, the leaf is elongated and narrow, if you pinch the leaf you can smell a very distinct odor, you pick the new growth and blossoms during the winter, dry I in the sun, bottle it in glass jars and keep it for use. To use this herb you take a pint jar, place a good pinch of the dried leaves in the jar then fill with boiling water, let cool for 24 hours, then drink it the following day and you also make another tea for the next day and you do this for 24 hours daily. The herb is bitter and can make you sick to your stomach which is a sure sign that you have cancer, you do not have to drink the pint jar all at one time but you must drink it that day. We had one of our friends who had cancer with us that winter so she was our guinea pig so to speak, so we collected the herb, dried it and then she started taking it, however she was feeling ill from it and didn't want anymore so her husband bought some empty capsules and we pulverized the herbs then filled the capsules and this helped her sickly feeling, she started taking the capsules about early February and when she got home in early April she went for a medical check up, and there wasn't any cancerous growths on her body, which she had been having removed for several years, and she continued to take the herb for 18 months then quit, however the native remedy called for 24 months to completely remove all the cancer.

The year we bought the King's Highway motor home we had it for the summer at our home in Surrey then took it south and parked it in our usual place on the slabs in Niland California. Although we had it built to suit we decided to make some changes on the interior of it, so we pulled the liquor cabinet out and something else and built a cupboard or two and also built in a pantry. While doing this to this nice looking motor home we had a lot of snowbirds coming around to see what we were doing to that new unit. We met one couple who had the same model and make that were from Grants Pass Oregon who later became good friends and bought some real estate together and spent a lot of time with them, we used to go fishing, traveled together for several years in fact it was Erma that we had cured the cancer for fourteen years.

After spending part of several winters on the Niland slabs near the Salton Sea, we started venturing out and spent several winters in

Mesa Arizona where we had a park model or I should say several park model homes and I also sold quite a few in that park. These parks all have the swimming pools, recreation centers, hobby shops, and a lot of snowbirds spend the whole winter in them; it is the winter home for many people who want warmer weather for their winter. Merelie really enjoyed staying there, but I was different and liked to move around and see more of the country, so always had a recreational vehicle of some kind. I would buy a motor home, check it over and add something to it or as the trade goes take the kinks out of it and sell. This was the seventies and eighties when the prices were always increasing so I was never out of pocket, between trailers and trucks, class C motor homes, and class A motor homes, I think we have our 18th unit this time in 2002.

We would usually come back to Cloverdale or later years to south Surrey and later to the Okanogan at Olalla. I sold real estate up to the late eighties when we found a nice piece of property in the Okanogan where we now call home, but I'm ahead of my story.

In the spring of 1978 Merelie and I returned to Cloverdale to take part in the activities of our horse club and also sell some real estate during the summer months. We were also trying to organize a pinto horse club for B.C. pinto horses which are white with black or red markings on each horse; generally speaking these horses can be from any breeding, however usually Arab, Thoroughbred, or Quarter horse but have a very gentle way with each horse and easy to train. We of course had a black and white pinto stallion named Chief Big Heart, he was of Arab and Saddle bred ancestry, a very knowledgeable horse. We had quite a few foals from him. He lived to 28 years then we had him put down when we sold our property in Cloverdale.

Selling real estate was quite interesting and easy for me. I would come back to B.C in the spring and start right off as if I had never been away for the winter; this was always the same each year until I gave my license up in 1991.

Merelie and I would go south each fall and spend our winter in California or Arizona. We lost interest in California when we didn't purchase the property in the Imperial Valley, although we have always liked that area, there was money to be made in Mesa Arizona and around that area; I could go in there and rent sites for park model mobile homes, have them set up and sell them quite easily, this was fun for me and I enjoyed my winters there. We had a nice park model in one of the parks in Mesa and Merelie really enjoyed being there as there was always lots of activities going on, water exercising, crafts of all types, always something to do. This carried on into the eighties, California and Arizona in the colder months and Surrey in the summer and sell real estate.

These activities and travel carried on until 1981 when we sold or traded our 1977 King's Highway motor home for another park model in a park in Mesa, and then we did a lot of selling and buying of recreational vehicles over the years.

In 1986 we sold our house and acreage in Cloverdale to be developed into an industrial park. We were not the only ones to get in on this deal. Some of our neighbors also sold for this development, one moved to Armstrong, another whom we had helped buy some property sold and bought a meat market in Vancouver, who later became quite wealthy. We bought a mobile home site in south Surrey near White Rock and bought a double wide mobile home and added another 12 feet to it for carport and workshop and a sun room. We lived there for several years and carried on traveling and selling as usual.

The fall of 1990 we took our motor home with Merelie's brother and sister-in-law on a tour of the wineries in the Okanogan. We traveled up through northern Washington on highway 2 and 20, to Osoyoos B.C. then on highway 97 through the Okanogan, called on several wineries then went on highway 3A over to Olalla to visit our friends there We stayed overnight and had breakfast with them then went for our walk through the little town, wondering why our friends had moved to this area. As it turned out we walked to the end of Main St. then started walking down each short street off Main St. and when we got to 7th St. our friend said "no use walking down there as there is only an old abandoned house there." Well Merelie said that we will walk down there anyway. We got down 7th St. part way, there was Olalla Creek running across the street and an old house sitting back in some bush, and all that could be seen was part of a green door. Merelie walked in on the property then called me to come and look. We had been looking for a piece of property with a creek on it for some time and here it was. We found out from our friends if she knew who owned the property and made an offer on it that day and as it turned out that is where our home is today.

Olalla was surveyed into lots in 1904 for a mining town where there are five abandoned gold mines in the immediate area. The surrounding area of the town site was owned by a man by the name of Hall whose wife still owned the 2 acres we bought in 1990; she was 91 years old and lived with her daughter in OK Falls. This property had been in a very colorful place with flowers and flowering trees, and at this time Mr. Hall had been dead for 17 years and the property had been vacant for 9 years. When we bought, it was overgrown with trees and bush although there were still some plants and apricot trees still bearing fruit, and there was also patches of flowers in the bush here and there.

When we bought the property and looked at the old house, I figured we would dismantle it, however after looking it over real good we decided it was too good to destroy and would remodel or revamp it. When we started to clean up and refinish the old house, while stripping one wall there was a note pasted on one board, "this house was built by Bert Erwin in 1922," and in 1990 Bert Erwin lived in Princeton B.C. and was in his early eighties, we had known him while we lived in the Princeton area. He told me that he had come into that area as a child carried on horse back from the Kamloops area, and when he grew up he became a carpenter and builder. We cleared up a lot of the brush before starting refurbishing the house so we would have room to work, and at one time we had bush piled up like hay stacks. We started work on the house in the spring of 1991 and moved into it in that late summer, there were two stone and cement chimneys which were lined with lumber also a fireplace and we were surprised to see the house still standing as we found places around the chimneys fireplace where the walls had been burning at sometime. There was also shavings and paper for insulation. There was lath and plaster on part of the walls and lumber and newspaper with wallpaper over it on the kitchen and back part of the house, the living room has a nine foot ceiling with tall windows, there are three bedrooms, kitchen, laundry room, front and rear porch, also a partial basement built with cement and stone, the floor joists are 3x6 planks rough but in excellent shape. We took all the lath and plaster off as it was starting to deteriate. While the walls were open we replaced the wiring and plumbing. We had to go into the attic to replace some of the wiring and found the attic full of vines that had grown up the walls and into the attic under the eves, it was unbelievable the amount of vines there. There was also dried corn cobs hanging on strings throughout the entire attic and after taking them from the attic it made two or three huge piles on the main floor. We then saw where the roof was leaking and had to replace the roof which had been cedar shakes. Keith had some time off work and came up from the coast to Olalla and helped us with the attic and another time helped us replace the roof. Up to now, Merelie and I had been doing all the work, which we enjoyed. Since I had given up my real estate license I had all the time I wanted to rebuild our new home. We replaced the lath and plaster with gyprock wallboard and painted these. We kept the original windows and doors as they were in good shape, had to rebuild the stairs outside front and back and also built a sun deck at the rear entrance. The floors are tongue and groove four inch lumber which was starting to show wear so we covered them with underlay and carpet except the kitchen and laundry room. Speaking of laundry room reminds me I must tell you about the laundry machine that was there when we bought, it was a real antique, it was set up in the garage and ran with the rear wheel of their car which was jacked up with a belt on it to operate the laundry machine, the water was pumped in from a well which was very good water, however the pipes and well pump had become rusty so we hooked onto the Olalla water system which is now costing too much so must clean up our well and go back on our own system, it is now February 2004 and time to do something about the water this year.

After buying this property we also bought a small farm tractor with a bucket on it, the make of it was Yanmar and was very handy for clearing the bush and also placing our roofing up to the roof while we did that work. We also moved the garage from it's original location to the present setting, rebuilt our driveway and many things that made our work easier.

In the late fall of 1990 we had been busy clearing up the property and were living in our recreation trailer, and one morning awoke to find we had about eight inches of light snow which meant it was time to leave and head south; that light snow made some colorful scenery along the creek, we took some pictures, put everything away and left that day, I think it was early November. At that time we had a 30 foot 5th wheel trailer and Ford pickup truck with propane for fuel which made a nice unit for Rving.

The next year we were using Keith's travel trailer as we had sold our truck and trailer and needed a place to live while fixing the old house. In the fall we had been falling some large cottonwood trees that were starting to rot and were dangerous to have on the property, In fact one large branch broke off one of the trees and came right through the garage roof. When we felled that tree part of it fell in the creek and another part of it fell on top of the old outhouse which I had planned on keeping for old time's sake, however it was flattened and we had to now move part of the tree out of the creek which I needed some help with. I asked a neighbor if he could help and he said yes as soon as he had time, but as it turned out a couple of weeks later it started to rain and now we had to move that tree to stop the creek from flooding us out; so the rain is really coming down hard and our neighbor came over and said that we better move that tree now, it is dark and raining and we went out with our tractor to move the tree, it took a couple of hours but we succeeded in getting it out of the creek. After cutting it in pieces we were both soaking wet, then went in the trailer, dried out and had a few drinks of wine and felt none the worse for the ordeal. These are things you don't forger because it should have been serious. Some of the black poplar was 2 feet across at the butt and over 200 feet tall, very tall for that type of tree.

The same fellow that helped me move the tree from the creek had an old car that was blocking the lane and was an eyesore where it was and we kept asking him to move it forward out of the way and closer to his house and after several weeks he moved it and shortly after that we had a freak heavy wind storm and the large tree beside where he parked his car came down in the opposite direction from the house and landed right on top of his car crushing it right to the ground. Across the road another large tree came down between two houses and none of the branches even touched either house. Either tree could have damaged or destroyed any of the houses.

This February 2004 Keith, my son and I are in our motor home in a recreation vehicle park near Guadalajara Mexico where we are spending some of our winter holidays. We have been traveling down the west coast of Mexico enjoying the ocean breezes, and now inland but shall soon return to the ocean areas for another month or so. In this park we have white pine and large trees that can accommodate eighty five recreational vehicles.

We have spent some time in Guadalajara Mexico looking at the sights and a large market, three storeys, I believe the largest in Mexico; we also drove out around Lake Chapala, the area is quite similar to the Okanogan in British Columbia, a lot of Canadians winter here.

Must get back to memoirs: After spending our winters south, we returned to Olalla. Keith was living in our house in south Surrey and it gave us a stopping place when in the area.

In 1992 we started building a larger new house on the opposite side of the creek. I should tell you about our property, as I said before we had been looking for a piece of property with a creek on it, and now we had two creeks, Keremeos Creek is on our east border and Olalla Creek runs through the center of the two acres and joins with Keremeos Creek on our southeast corner.

The new house would be on 7th Street surrounded by the creeks on three sides, a very nice location. This Olalla Valley has very little wind, a moderate climate year round and sunshine almost every day.

We did find one problem, when we dug out the basement we had to stay above the creek level or the water would seep in, so this meant the house had to be built much higher if we wanted a basement, we put in a partial basement and of course the main floor of the house is about 5 feet above ground level, about 8 feet of stairs, this was alright for a while but Merelie said the stairs had to go or something else had to be done.

In 1997 Merelie and I and some of my family went to Ontario for part of the summer, and when we returned I proceeded to build a suite under the house, and completed it in October and moved in from the top storey. I bought a small tractor with a bucket so we could go under the house to dig out the dirt and part of the basement, we now have a nice apartment, cool room for vegetables, frig and freezer room, also a wine room for making our own wine, also a 6 foot walkway down to the basement level, no stairs and Merelie was quite pleased.

Merelie, My wife and best partner passed away in July 2000, she had been ill for some time after remodeling the old house and demolishing old buildings including a greenhouse. Apparently she got some poisons in her lungs, pesticides, fungus, mildew, which started her illness, pulmonary fibrosis, a growth in the lungs which eventually fills the lungs and kills you. There is a cure for it which our doctors are not aware of, however the naturopath doctor will use a curing remedy used in Europe, but we didn't know soon enough. Merelie and I would have been married 60 years in the following November.

After Merelie passed on I didn't do any traveling for two years until my son Keith suggested I go south with him for the winter. We went to Texas for two months and the following year we went to Florida and the last two years we went to Mexico.

Before Keith, my son retired in October 2002 I decided to sell the house in Surrey that Keith was living in and he would come and live with me in Olalla. As it turned out, the house sold before Keith retired so he parked the motor home we had recently bought and lived in it until October, then he moved up to Olalla in the Okanogan with me.

I started looking for another house, so I bought a house to remodel to put back on the market. The carpenter I had did not get it finished so we had to hire another carpenter to finish it, rented it out until summer then put it up for sale, when it sold.

I have bought and refurbished a number of homes over the years, it usually is profitable, however that last one was not and it may have taught me to be more cautious, I have always been interested in real estate and know there is a profit to be made but try to estimate your cost before buying.

Over the years I have had some that I had to foreclose on because of lack of payment but this is part of the things you have to put up with. I have been fortunate in usually making a profit, have one that is behind thirty payments which is going through the court at this time but thanks to our Lord we shall succeed in full payment. I'm getting up in years but this type of dealing keeps one's mind active and you don't become a couch potato.

I have enjoyed everything I have done over the years, never found or tried anything that wasn't interesting. Had a good marriage and family, what more could we ask for? I'm still looking forward to RVing, like going south to California or Mexico, walking the ocean beaches and exploring to see what other people are doing; if you look and enquire you will find very interesting people as you travel.

Into this book goes over eighty years of my life history, so I know I have tried and accomplished many things; this is most of them and the experiences I have had.

I suppose a lot of my experiences and accomplishments have come down from both sides of my ancestry, as each had accomplishments that some have been recorded but many more should have been.

I believe you, the reader will find this book very interesting reading and it is all true to the best of my recollection.

I'm retired in a very nice area of the Okanogan Valley, two acres of ground with 2 creeks; the Olalla Creek runs through the middle of these two acres and joins up with the Keremeos Creek on the east border. I have a two very nice apartment home with my daughter and her family on the top floor and my son and I on the bottom apartment.

My daughter is an exceptionally good landscaper; we are blessed with lots of flowers, ornamental trees, also fruit trees, and a very productive vegetable garden.

The following is a partial list of things that I have done throughout my lifetime as they appeared in the book:

farm boy purchasing agent
magazine salesman bush contractor
road laborer farmer
distributing trout from hatchery gas station operator
bush worker oil company maintenance supervisor
pulp cutter theatre maintenance & doorman
scaler sanitary supply sales man
cruiser resort & ranch operator
camp clerk real estate salesman & investor
shipyard carpenter
house and heavy construction carpenter
dairy operator

ISBN 141209720-7